Build Your Own 386/386SX Compatible and Save a Bundle
2nd Edition

Aubrey Pilgrim

Windcrest®/McGraw-Hill

SECOND EDITION
FIRST PRINTING

© 1992 by **Aubrey Pilgrim**.
First edition © 1988 by TAB Books.
Published by Windcrest Books, an imprint of TAB Books.
TAB Books is a division of McGraw-Hill, Inc.
The name ''Windcrest'' is a registered trademark of TAB Books.

Printed in the United States of America. All rights reserved. The publisher takes no responsibility for the use of any of the materials or methods described in this book, nor for the products thereof.

Library of Congress Cataloging-in-Publication Data

Pilgrim, Aubrey.
    Build your own 386/386SX compatible and save a bundle / by Aubrey Pilgrim—2nd ed.
      p.   cm.
    Includes index.
    ISBN 0-8306-3752-4   ISBN 0-8306-3750-8 (pbk.).
    1. Microcomputers—Amateurs' manuals. 2. Intel 80386 (Microprocessor)—Amateurs' manuals. 3. Intel 80386 SX (Microprocessor)—Amateurs' manuals. 4. IBM-compatible computers--Design and construction—Amateurs' manuals. I. Title.
TK9969.P54 1992
621.39'16—dc20   91-35087
                                                                           CIP

TAB Books offers software for sale. For information and a catalog, please contact TAB Software Department, Blue Ridge Summit, PA 17294-0850.

Director of Acquisitions: Ron Powers
Book Editor: David M. McCandless
Production: Katherine G. Brown
Book Design: Jaclyn J. Boone
Cover: Sandra Blair Design, Harrisburg, PA   EL1

Contents

Introduction *xiii*

1 Why you need a computer *1*
   Why you need a computer *2*
   To build or buy *2*
   To build or upgrade *3*
   Why it costs less to build or upgrade *3*
   386 vs. other systems *4*
   What the 386 can do *5*
   Real and protected modes *5*
      Real mode *5*
      Protected mode *6*
   Do you need a 386? *6*
   Sources *6*

2 Necessary components *7*
   386SX and 386DX motherboards *9*
      Cache *10*
      Coprocessors *10*
      ISA and EISA *12*
         The Industry Standard Architecture (ISA) *12*
         EISA and the Gang of Nine *13*
         The EISA connector *13*
         Do you need an EISA computer? *14*
      Cost of basic components *14*
      Case *15*
      Switch panel wires *15*

Power supply *16*
Uninterruptible power supply *16*
RAM memory sockets *18*
Monitor *18*
Monitor adapter *18*
Multifunction boards *18*
Floppy drives *19*
Hard disks *19*
Disk controllers *19*
Keyboards *19*
Where to find the parts *19*

3 386SX or 386DX assembly *21*

Benchtop assembly *22*
    Assembly instructions *23*
Installing the components in the case *36*
    Switch panel wires *40*
TEST.BAT *40*
Slot covers *41*

4 Upgrading your computer *43*

Put new life in your old clunker *44*
    Baby-size motherboards *44*
Cost to convert *44*
    Case *44*
    Power supply *45*
    Keyboard *45*
The conversion of an IBM XT *45*
Accelerator boards *54*
CPU daughter boards *55*

5 Floppy disk drives *57*

Some floppy disk basics *58*
    Tracks *59*
    Formatting *59*
    Sectors *59*
    Allocation units *59*
    FAT *59*
    Directory limitations *60*
    Cylinders *60*
    Heads *61*
    TPI *61*
    Read accuracy *61*

Rotation speed *62*
The 5.25" standard *62*
Formatting *65*
    Helpful .BAT files *66*
    Converting a 720K to a 1.44M disk *66*
    Cost of disks *67*
    Discount disk sources *68*
Floppy controllers *68*
Higher density systems *68*
    Extended density drives *68*
    Very high density drives *68*
    Bernoulli drives *69*
Mounting a 3.5" drive *69*
What to buy *69*

6 Hard disks and mass storage *71*

Choosing a hard disk *72*
    Capacity *73*
    Speed or access time *73*
    Type of drive, stepper, or voice coil *73*
A bit of history *73*
What size disk do you need? *74*
How a hard disk operates *74*
    Leaving it turned on *74*
    MTBF *75*
    Reading and writing *75*
    Platters *75*
    Head positioners *76*
Drive systems *76*
    MFM *76*
    RLL *77*
    IDE *77*
    SCSI *78*
    ESDI *78*
    Controllers *78*
Formatting a hard disk *79*
CMOS ROM setup *80*
Adding one or more hard drives *80*
Hard cards *80*
External drives *81*
Compression *81*
CD-ROM *83*
    The High Sierra standard *84*

7 Backups *87*

Unerase software *88*
Jumbled FAT *89*
Head crash *89*
Crash recovery *90*
    Small logical drives are better *90*
Excuses *90*
    "I don't have the time" *91*
    "It's too much trouble" *91*
    "I don't have the necessary disks, software, or tools" *91*
    "Failures and disasters only happen to other people" *92*
Additional reasons to backup *92*
    General failure *92*
    Theft and burglary *92*
    Archival *92*
    Fragmentation *92*
    Data transfer *93*
Methods of backup *93*
    Software *93*
    BACKUP.COM *93*
    Tape *94*
    DAT *95*
    Videotape *96*
    Very high density disk drives *96*
    Second hard disk *96*
    External plug-in hard drives *97*
    Hard cards *97*

8 Monitors *99*

What I recommend *100*
    A checklist *101*
Monitor basics *101*
    Scan rates *102*
    Controlling the beam *103*
    Monochrome *103*
    Color *103*
    Resolution *104*
    Pixels *104*
The need for drivers and high resolution programs *105*
    Interlaced vs. non-interlaced *105*
    Dot pitch *106*
    Bandwidth *106*
    Landscape vs. portrait *106*
    Screen size *106*
    Controls *107*

　　　　　Glare   *107*
　　　　　Cleaning the screens   *107*
　　　　　Tilt-and-swivel base   *107*
　　　　　Cables   *108*
　　Adapter basics   *108*
　　Analog vs. digital   *110*
　　Very high resolution graphics adapters   *110*
　　Drivers   *110*
　　Monitor and adapter sources   *111*
　　　　　List price vs. street price   *111*
　　What to buy if you can afford it   *111*

9　Memory   *113*

　　A brief explanation of memory   *115*
　　DRAM   *115*
　　　　　Refreshment and wait states   *115*
　　　　　Interleaved memory   *116*
　　　　　Cache memory   *116*
　　　　　SRAM   *116*
　　　　　Flash memory   *117*
　　Motherboard memory   *117*
　　　　　Dual in-line package (DIP)   *117*
　　　　　Single in-line memory module (SIMM)   *117*
　　　　　Single in-line package (SIP)   *117*
　　How much memory do you need?   *119*
　　Types of memory   *119*
　　　　　Conventional memory   *119*
　　　　　Extended memory   *119*
　　　　　Expanded memory   *120*
　　Memory modes   *120*
　　　　　Real mode   *120*
　　　　　Standard or protected mode   *121*
　　　　　The 386 enhanced mode   *122*

10　Input devices   *123*

　　Keyboard covers   *124*
　　A need for standards   *124*
　　Model switch   *126*
　　How a keyboard works   *126*
　　　　　Special keys   *126*
　　　　　Function keys   *126*
　　　　　DOS and function keys   *127*
　　　　　Numbers   *127*
　　　　　Arrow keys   *127*
　　　　　Home and end keys   *128*

　　　　PgDn and PgUp (Page down and page up)　*128*
　　　　Ins (insert)　*128*
　　　　Del (Delete)　*128*
　　　　Esc (Escape)　*128*
　　　　PrtSc (Print screen)　*128*
　　　　* (asterisk)　*128*
　　　　Scroll lock　*128*
　　　　Break　*129*
　　　　Backspace　*129*
　　　　Return or enter　*129*
　　　　Shift　*129*
　　　　CapsLock　*129*
　　　　Ctrl (control)　*129*
　　　　Tab　*129*
　　　　Alt (Alternate)　*130*
　　　　Other special key functions　*130*
　　Reprogramming key functions　*130*
　　Keyboard sources　*130*
　　Specialized keyboards　*131*
　　Mouse systems　*132*
　　Mouse types　*132*
　　Mouse interfaces　*133*
　　Mouse cost　*134*
　　Trackballs　*134*
　　Keyboard/trackball combination　*134*
　　Digitizers and graphics tablets　*134*
　　Scanners and optical character readers　*135*
　　　　OCR　*135*
　　　　Hand-held　*135*

11　Printers　*137*

　　Choosing a printer　*138*
　　　　Dot matrix printers　*138*
　　　　Dot matrix speed　*139*
　　　　Dot matrix color　*139*
　　　　Advantages of dot matrix　*139*
　　　　Noise reduction　*139*
　　　　Low cost　*140*
　　　　Print buffer　*140*
　　　　Daisy wheel printers　*140*
　　　　Ink jet　*140*
　　　　Ink jet color　*141*
　　　　Laser printers　*141*
　　　　Engine　*141*
　　　　Low-cost laser printers　*142*

        Extras for lasers   *143*
        Memory   *143*
        Page description languages   *143*
        Laser speed   *144*
        PostScript printers   *144*
        PostScript on disk   *144*
        Resolution   *144*
        Paper size   *145*
        Maintenance   *145*
        Paper   *145*
        Address labels   *145*
Color   *146*
Plotters   *146*
        Plotter supplies   *147*
Installing a printer or plotter   *147*
Printer sharing   *148*
Sources   *149*

12   Telecommunications   *151*

Basic types of modems   *152*
        External   *152*
        Internal   *152*
Communications software   *152*
Protocols   *153*
Baud rate   *153*
How to estimate connect time   *154*
What to buy   *154*
Installing a modem   *155*
        Set configuration   *155*
        Plug-in telephone line   *155*
        A simple modem test   *156*
        Cables   *156*
Bulletin boards   *156*
Viruses and Trojan horses   *157*
Illegal activities   *157*
Cost to use   *157*
Where to find the bulletin boards   *158*
On-line services   *158*
E-mail   *158*
Banking by modem   *159*
Public domain and shareware   *159*
ISDN   *159*
Facsimile boards and machines   *159*
        Standalone FAX units   *160*
        FAX computer boards   *160*

Single board modem and FAX *161*
Installing a FAX board *162*
Telecommuting *162*
More help *162*

13 Windows *163*

A need for standards *164*
Windows 3.0 *164*
GUI *164*
    Why Windows is easier to learn and use *165*
    Requirements *165*
    Automatic setup and installation *165*
    On-line help *165*
    Breaking the 640K barrier *166*
    Operational modes *166*
Other features of windows *166*
    Dynamic data exchange (DDE) *166*
    Calculator *166*
    Calendar *166*
    Cardfile *166*
    Clock *166*
    Notepad *167*
    Recorder *167*
    Clipboard *167*
    Write *167*
    Paintbrush *167*
    Terminal *167*
    Reversi and Solitaire *167*
    Wallpaper *168*
    Windows vs. OS/2 *168*
Applications for windows *168*
    YourWay *168*
    Norton Desktop for windows *169*
    Micrografx *169*
    Ventura Publisher *169*
    PageMaker *169*
    Crosstalk for Windows *169*
    Windows Express *169*
    Dragnet and Prompt *169*
    Word processors *170*
GeoWorks *170*
New windows programs *170*

14 Essential software *171*

Operating systems software *172*

MS-DOS 5.0 *172*
DR DOS 5.0 *173*
OS/2 *173*
DESQview *174*
Concurrent DOS 386 *174*
DOS help programs *174*
Word processors *174*
    WordStar *174*
    WordPerfect *175*
    Microsoft Word for windows *175*
    AMI *175*
    PC-Write *175*
Grammar checkers *175*
Database programs *176*
    dBASE IV *176*
    askSam *176*
    R:BASE 3.1 *176*
    FoxPro *177*
    Paradox *177*
Spreadsheets *177*
    Microsoft Excel *178*
    Quattro *178*
    SuperCalc5 *178*
Utilities *178*
    Norton Utilities *179*
    Mace Utilities *179*
    PC Tools *179*
    SpinRite II *179*
    Disk Technician *179*
    OPTune *179*
    CheckIt *179*
    SideKick Plus *180*
Directory and disk management programs *180*
    XTreePro Gold *180*
    QDOS 3 *180*
    Tree86 3.0 *180*
    Wonder Plus 3.08 *180*
Search utilities *180*
    Magellan 2.0 *181*
Computer Aided Design (CAD) *181*
    AutoCAD *181*
    DesignCAD 2D and DesignCAD 3D *181*
Tax programs *181*
    Andrew Tobias TaxCut *182*
    J.K. Lasser's Your Income Tax *182*
    SwifTax *182*

TaxView *182*
TurboTax *183*
Miscellaneous *183*
Money Counts *183*
It's Legal *183*
WillMaker 4.0 *183*
Random House Encyclopedia *183*
ACT! *184*
Form Express *184*
Summary *184*

15 Mail order and magazines **185**

Mail order *186*
Ten rules for ordering by mail *187*
Federal Trade Commission rules *188*
Computer magazines *189*
Free magazines to qualified subscribers *190*
FaxBack *191*
Public domain software *193*
Mail-order books *194*

16 Troubleshooting **195**

Fewer bugs today *196*
Document the problem and write it down *196*
Levels of troubleshooting *197*
Electricity—the life blood of the computer *197*
The basic components of a computer *198*
Power On Self Test (POST) *199*
The impossible quest *200*
Power supply *200*
Instruments and tools *201*
Common problems *202*
Electrostatic discharge (ESD) *202*
Recommended tools *202*
How to find the problem *203*
Diagnostic and utility software *204*
What to do if it is completely dead *205*
Software problems *206*
Hardware problems *207*
Is it worth repairing? *208*

Glossary **209**

Index **223**

Introduction

This book will show you how easily you can build a fast and powerful 386SX or 386DX. It contains a great number of photographs and easy-to-understand instructions. It describes the necessary components and their operations in detail and and tells how they interact to make up a system. Thus, even if you are not building a computer, this book can help you understand how a computer operates. If you are in the market for a computer (and everyone should be, in my opinion), this book can definitely help you save some money.

Some of you might have doubts about being able to build your own computer. Actually, the word *build* is not a good term for this book; the term *assemble* would be much more accurate. You don't have to do any soldering or wiring—the components you buy (such as boards, disk drives and other peripherals) have already been assembled, soldered, and tested. You merely choose the components that you want to put into your computer. The components—the floppy disk drives, the hard disk drives, the monitors, keyboards, and other components (from various vendors)—are all interchangeable. So you can shop around to find the best buys, and then merely connect and assemble them into a system. Believe me, this process is very simple. After you have bought the components, it will only take an hour or so to put your system together.

Anyone can do it

These computers might be very sophisticated and advanced, but you shouldn't have any trouble assembling one. You don't need any special expertise, technical knowledge, or special skill; and no soldering or special tools/instruments are required. You merely need just a pair of pliers and a screwdriver. Anyone can do it.

You will probably have more trouble running some of the so-called "user friendly" software than in assembling a system. I have been using computers for over 15 years, but I still have great difficulty using some software, especially when trying to understand the manuals and documentation.

Industry changes

The computer industry has changed greatly since I wrote the original 386 book. For example, when I built my first 386 in 1988, I paid $1825 for the motherboard alone and the board operated at 16MHz (the original standard frequency). Today I can buy a smaller, faster, and better motherboard for about $400.

Until mid-1991, Intel was the sole manufacturer of the 386 CPU. Recently, American Micro Devices (AMD) was given the right to manufacture the 386. The Chips and Technology has also announced a cloned 386 CPU. For the first time, Intel has a bit of competition, which will drop the prices even more. The prices of other components have also come down. Even with inflation, you can build a much bigger, better, and more powerful computer for less money than it would have cost a few years ago. Wouldn't it be great if the prices of automobiles and the many essential luxury items would follow the same downward trend?

Differences between 386SX and 386DX

The economical 386SX can do everything that the powerful 386DX can do, except for the fact that it's a bit slower. The 386DX is a true 32-bit CPU, handling data externally and internally 32 bits at a time. The 386SX handles data externally over 16-bit lines; but internally, it processes the data in 32-bit chunks, just like the 386DX. Except for the motherboard, both 386 machines use the same components.

Again, except for the motherboards, the components needed to assemble a 386SX or DX are basically the same as those used in the XTs, 286s, and 486s. The machines differ in operating speed, though. The Intel 386SX operates at 16 or 20MHz, while the AMD operates at 25MHz. The Intel 386DX presently operates at 25 or 33MHz, while the AMD 386DX operates at 40MHz.

Cost to build

Just a short time ago, the 386 was much too expensive as a personal computer for the homeowner. It was even too expensive for some businesses. The IBM PS/2 Model 80 still costs up to $10,000 or more. Many of the other name brand computers such as Compaq and Macintosh might cost about the same.

However, you can assemble a good 386 system that will do everything that the IBM, Compaq, or Macintosh will do for as little as $1000 and up to $3000. Many available non-brand name clones are very competitively priced. Also, the magazines are full of ads for very low priced machines.

If you compare the prices of these machines and the prices of the individual components needed to build a machine, it might appear that you could not save much by doing it yourself. Read the ads closely, though; some are quite misleading. They might have small print that says "without monitor, w/o CPU, or 0K." If you don't want to assemble one yourself, you can buy a "bare-bones" system and add to it.

This book will help you make informed decisions on what to buy. It will help you save time and a bundle of money whether you build your own or buy it. How much you pay for your computer will depend on how well you shop, where you buy the components, and what components you put in your computer. You might not save a whole lot compared to some of the no-name clones, but you will be able to save two or three times the cost of an IBM, Compaq, or Macintosh.

Applications for a 386

Schools

The 386 can be used in thousands of applications and ways in a school setting, offering a great advantage to students and teachers. Software such as spelling checkers, word processors, and graph programs can be used in a variety of curriculums from English to Physics. When I taught a high school class in the early 1980s, a computer would have made it so much easier for me to organize and to instruct; unfortunately, I couldn't afford one then. Because schools never seem to have enough money to afford computers, perhaps purchasing the relatively cheap components and allowing the students to assemble them would be both economical and also a great learning experience.

Churches, clubs, and fraternal organizations

Churches also never seem to have enough money. The pastor could use a computer to write his sermons, keep a list of his parishioners, keep track of how much money is needed to get a new roof, and help with other financial accounting, desktop publishing, etc. In addition, clubs and fraternal organizations must keep track of their members, required dues, and financial records. They also need address and mailing lists, as well as simple desktop publishing to send out letters announcing planned events. A church or club member could easily assemble a computer for their organization.

Businesses

Computers are essential in all businesses—large and small—for obvious reasons. A large business could have one or more of their employees assemble the computers and save large amounts of money. Small business people would also save money by assembling their own computers. With both the reduced price and the power of the computer, the small business might even save enough money to expand into a big business someday.

Doctors and lawyers

Computers today are as essential to doctors and lawyers as a stethoscope and a briefcase. Doctors can use a computer to keep track of patients, case histories, and billing. They could also use computers to keep track of their business portfolio and investments. Similarly, lawyers can use a computer to keep a list of their clients as well as the time spent on their behalf doing research, making out wills/contracts/agreements, writing threatening letters, and doing other types of lawyer-type activities.

Home and home office

A 386 can be used for work brought home from the office, to manage business investments and portfolios, to create and manage a budget, to save addresses, to list all of your possessions, to write nasty letters to the editor, to aid a small home office business, plus many more things. In addition, many of the newer homes will be set up so that a computer can control the lighting, heating, and other functions.

Other uses

I can't possibly list all of the uses for a 386. In addition to the professions already mentioned, it can be used by

- realtors.
- bankers.
- farmers.
- architects.
- design engineers.
- manufacturers and managers using Computer Aided Design/Computer Aided Manufacturing/Computer Integrated Manufacturing/Computer Aided Engineering/Computer Aided Software Engineering (CAD/CAM/CIM/CAE/CASE).
- desktop publishers.
- network file server administrators.
- hundreds of others.

Chapter 16 discusses some of the business uses for your computer. I have included a few short reviews of some hardware and software and offer a few suggestions for certain applications. One great advantage of the 386 system is that it can be configured in almost any way that you want. In order for you to make informed choices, this book describes in detail the operation of some of the peripherals needed (such as the disk drives, the monitors, and printers). I also go into some detail and list a few of the hundreds of applications for which the 386 system can be used. I recommend some standard off-the-shelf software and explore such things as desk top publishing, networks, and other business and personal uses for the 386.

How compatible

IBM created a *de facto* standard for the PC, XT, and AT. This standard was easily copied and duplicated by the compatible clone manufacturers. Soon, about five billion dollars worth of hardware and about six billion dollars worth of software for the PC/MS-DOS system were being marketed. The compatibles and clones were sold at much lower prices, resulting in IBM losing a large chunk of the market.

IBM tried to recapture that market by developing the PS/2 MCA systems, which are incompatible with all earlier IBM and clone hardware. The PS/2 systems are able to use most of the six billion dollars worth of software that has been created but not any of the inexpensive boards and much of the five billion dollars worth of hardware that had been developed for the IBM and compatible systems. If you bought a PS/2 from IBM, you had to buy almost all of the optional hardware from IBM. This didn't create too much of a problem for large corporations that didn't worry too much about budgets, but it was very expensive for the ordinary person.

You cannot justify paying IBM prices when you can get an equivalent system for about one-third less the price. The clone systems have no problem using existing hardware. Over 40 million of the older IBMs and compatible clone systems exist; you can be sure that much more hardware and software will continue to be developed for these systems.

Upgrading an older PC, XT, or AT

I'll show how to upgrade and transform an older PC, XT, or 286 AT machine into a fast and powerful 386 by simply installing a new motherboard or by using one of the 386 plug-in accelerator boards. The procedure takes only a few minutes and is relatively inexpensive. Photos and detailed instructions I've included will help you do this. That chapter alone could save a person hundreds or even thousands of dollars, preventing obsolescence and protecting valuable investments.

I wrote this book in a language that should be easily understood by the novice, yet is technical enough to satisfy the experienced user. The subjects of each chapter as well should be of interest to the beginner as well as to the old pro.

Chapter contents

Here's a run-down of what this book contains:

Chapter 1 describes the 386 and gives reasons why a person should build their own.

Chapter 2 contains descriptions of parts and components needed to assemble a 386.

Chapter 3 contains photos and instructions on how to assemble a 386.

Chapter 4 describes how to upgrade a PC, XT or 286 to a 386SX or 386DX. Or even to a 486SX or 486DX.

Chapter 5 describes operations of floppy disks.

Chapter 6 describes the operations of hard disks.

Chapter 7 lists reasons why you should backup, as well as some methods for doing so.

Chapter 8 describes monitors in detail.

Chapter 9 describes the function and operation of memory.

Chapter 10 talks about keyboards and other methods of input to the computer.

Chapter 11 discusses the different types and features of printers.

Chapter 12 discusses communications software and hardware, including modems, E-mail and FAX.

Chapter 13 briefly discusses Windows and some of the software that run under it.

Chapter 14 reviews some of the essential software that you will need.

Chapter 15 discusses sources for your components. I also discuss mail orders and give a list of computer magazines, some of which are entirely free.

Chapter 16 has a few tips on troubleshooting in case something goes wrong.

The glossary contains hundreds of new buzz words and acronyms invented for the computer revolution. This extensive list should help you hold your own in any computer setting.

Writing this book (or any other) was not an easy thing to do because it demanded long hours and much research. Also, at times, it practically isolated me from my family, friends, and loved ones. Still, writing has its rewards—the very act itself, as well as the resulting sense of accomplishment, can be most satisfying. Of course, if I sell a few books, it helps even more.

Thank you for buying this book. I hope you enjoy it, and I hope that you also save a bundle by building your own computer.

Chapter 1

Why you need a computer

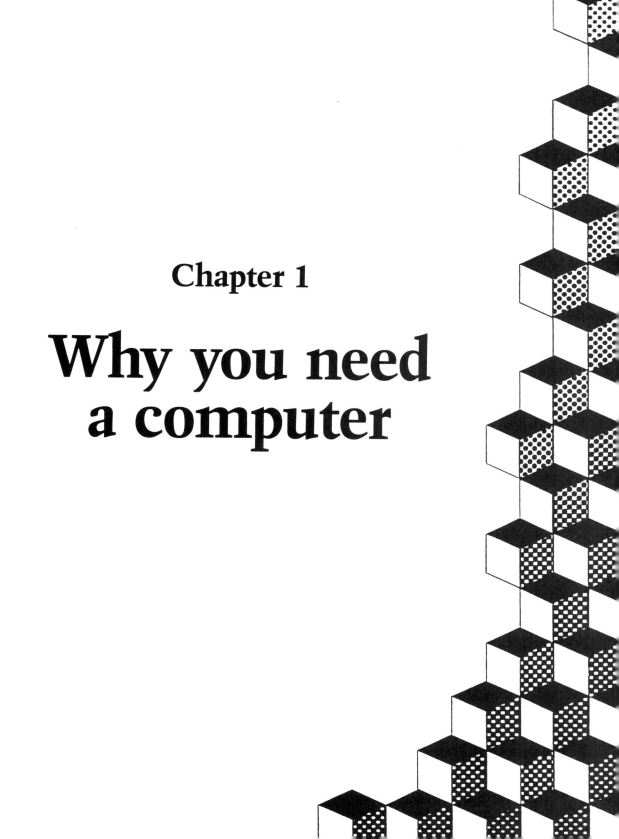

Several distinct ages have been designated throughout history, such as the stone age, the bronze age, the iron age, and now the silicon (or computer) age. It is rather ironic that early man wrote on stone and now we're writing on silicon, which is also a form of stone.

Technology is changing faster today than at any time in history, basically due to the computer and its ever-changing technology. One can hardly keep up with the latest advances; we are continuously inundated with information about new technologies, new products, new research findings, and new developments. I subscribe to over 50 computer magazines, many of which are issued weekly (such as InfoWorld and PC Week) or bi-weekly (such as PC Magazine). Incredibly enough, every issue of each magazine has something new in it.

Why you need a computer

Computers are here to stay, having become an essential part of our life. If you have no special skills, you might be competing with thousands of others for a menial job that doesn't pay very much. This world is pretty tough to live in sometimes. We must seize every advantage that we can to get ahead or, in some cases, just to survive. *Carpe diem* is a Latin phrase that literally means to "Seize the day" (or loosely, "Get what you can while you can"). If you don't own a computer, you might be at a disadvantage and not be able to get all that you can.

Children especially need to learn computers because, as adults, they will be even more dependent on computers than we are today. If you have children, you might feel guilty about not having a computer for them, and well you should. You might have delayed buying a computer because of the high prices, but you can now build one for a very reasonable cost.

To build or buy

You might have some reservations about being able to build your own computer, but you shouldn't feel that way. You can do it; anyone can.

You might have seen ads for very low-cost systems that suggest that it would cost as much to build a system as to buy an already assembled one. Still, you must read the ads closely. I've seen ads for the Apple Macintosh for a very low price; but in small print, the ads might say that the price doesn't include a monitor, keyboard, or possibly a hard drive.

You must also determine whether the low-cost advertised system is equivalent to the system you want. Many options are available, with how much you save depending on how you shop. An equivalent clone component might only cost half as much as a brand name one, but the components are all interchangeable. Look through the ads in the computer magazines; a large difference in a component's price might exist between different vendors.

If you happen to be on a budget, you can buy a few parts at a time and gradually build your computer up. As I mentioned earlier, most of the components are the same as those used in the PCs, XTs, and ATs, except for the motherboard.

If you are still hesitant about building your own system from scratch, you can buy a bare-bones model with just the basic components and then add to it. Although it is quite easy to assemble a system, you could make a mistake or get a defective component. The bare-bones systems have usually been checked out, so they could save you some time and problems.

I would recommend that you build your own computer; you'll learn a great deal and feel good too.

To build or upgrade

If you already have an older system that does everything you need it to do, you might not want to invest in a shiny new 386 computer. If you have more time than money, you might be able to wait a few milliseconds for your old machine to accomplish a task. Maybe you don't need new multitasking, multi-using, and very fast software packages if your old programs are doing the job; maybe you can get by with just a few upgraded pieces of hardware and some new software. In Chapter 4, I show that you can upgrade your old computer to make it as good as a brand new one.

Possibly, you might give your older system to the kids or to your secretary and build a new fast and powerful 386. If you decide that you do need a new computer system, this book can help you make an informed decision. In the old days, you could choose fairly easily between which computer and peripherals to buy; you didn't have many choices. Today, so many choices exist that trying to determine which products to buy can be a dilemma. If you're faced with this problem, this book can help you.

Why it costs less to build or upgrade

One advertising campaign some time ago said that no one had ever been fired for buying an IBM. Considering the high cost of IBM products compared to the clones, that statement might no longer be true. An IBM 386 Model 80, depending on the configuration, might cost from $9,000 up to $12,000. An equivalent clone system that will do everything the IBM will do can be assembled for about one third of the price of an IBM system.

One reason that the IBMs, Compaq, Macintosh, and other brand name computers cost so much is that they are sold through distributors. Several distributors and middlemen might handle the products from the time they leave the factory until they end up on the showroom floor, and each person who handles the product makes a profit. The large companies usually have fancy showrooms, several salespersons, and a ton of other overhead to pay for. On the other hand, most of the clones have few or no distributors and middlemen between the factory and the outlet. Many of them sell by direct mail and don't even have a showroom to worry about. Many of these cost savings are then passed on to the customer. In addition, the existence of so many clone vendors helps to drive the prices down.

4 *Why you need a computer*

386 vs. other systems

From the outside, a 386 machine might look very much like an AT, XT, or even a plain old PC. The main difference is that the heart of the 386 machine is an 80386 microprocessor chip. When enclosed in a ceramic package with all the leads and pins attached, the package is about 1.5″ square. Inside the package, however, is a small .5″ square of silicon with over 275,000 transistors etched into it. Figure 1-1 shows a 386 chip alongside a penny.

1-1 A 386 chip with 275,000 transistors compared to a penny.

Actual chip
Actual size

The chip shown is the original version. The newer version of this chip (see Fig. 1-2) is about one third smaller, at only .275″ × .275″ square. The 8088 microprocessor used in the PCs and XTs had about 29,000 transistors, and the 80286 chip has about 130,000 transistors on a small .5″ square piece of silicon. The 486 chip has 1,200,000 transistors on a piece of silicon .4″ wide by .65″ long.

During manufacture, one or more of those 275,000 transistors on a 386 chip could end up being defective. The larger the number of transistors on a chip, the greater the odds of defects occurring. The chips are tested, and the defective ones

1-2 A 486 chip with 1,200,000 transistors compared to a penny.

are rejected. Of course, the rejects reduce the yield. As a result, the 386 is rather costly compared to 8088 and 286 microprocessors. Up until mid-1991, Intel was the only manufacturer of the 386. Now American Micro Devices (AMD) and Chips and Technology are also making the 386, so we now have some competition. This clone chip is faster and uses less power. The prices of 386 CPUs are now much more reasonable.

The original 386 chip was designed to operate at 16MHz. They now operate at 25, 33, and 40MHz. By the time you read this book, some will be operating at 50 and 66MHz. Of course, the higher the operating frequency, the more expensive the CPU and motherboard.

What the 386 can do

A minicomputer or a workstation can cost as much as $50,000. For some applications, a 386 system costing between $3,000 to $10,000 can do an equivalent job. In many cases, the minicomputers and the workstations can only use special and very high priced custom software.

About 7 billion dollars worth of existing software was developed for PCs, XTs and ATs. The 386 can use all of it. The 386 will be even more powerful and versatile as software continues to be developed to utilize more of its fantastic capabilities.

One of the more laudable features of the 386 is its speed. Computers use quartz crystal oscillators to mark very precise blocks of time in which to perform an operation. The standard 8-bit 8088 CPU operates at 4.77MHz. It takes 8 bits to make a single alphabetic character or numeral. The 32-bit 386 can operate as high as 25, 33, and 40MHz, thus handling four times as much data ten times faster than the standard 8088. Actually, it handles data more efficiently than the 8088, so the speed of operation is even faster.

A lot of software (such as large spreadsheets, databases, or computer aided design programs—such as CAD), can require a long time to process. Waiting for some of these programs to run on an XT or 286 might cause you to drop off to sleep. Waiting even a few seconds for a computer to perform an operation can seem like an eternity. Saving a few seconds might not seem like much; but if the computer is used for long periods of time, the savings can more than pay for the extra cost of the 80386.

Real and protected modes
Real mode
When the computer is reset or booted up, it starts in the real mode. In this mode, it acts like a very fast 8088 or 286-based computer. This is the mode that we have been using for years. Even if you have 64M of RAM memory, unless the software is designed to take advantage of the extra memory, you will be limited to the real mode 640K.

Protected mode

With the appropriate software or operating system, the 386 can be switched to the protected mode. While operating in the protected mode, it can do multitasking, which allows the user to work on two or more tasks at the same time. Thus, you could have a window on half of the screen with a word processor running and a database program on the other half. Data from either file could be transferred from one window to the other.

In the protected mode, a type of barrier is imposed between programs being run concurrently so that neither one can interfere with the other. Without the barrier of the protected mode, loading two programs into a computer at the same time would cause them to be inextricably intermixed in the DRAM memory, comparable to pouring a gallon of hot water and a gallon of cold water into a barrel and then trying to separate them.

The early microcomputers could only address 64K of memory; later, the PC and the XT were able to address 640K. With the proper software, the 286 can directly address 16M of memory. The 386 can address 4 gigabytes with the proper software.

Do you need a 386?

The power and superb capabilities of the 386 make it well suited for business uses. If you use a computer just to do word processing and maybe a small spreadsheet once in a while, then you can get by with a PC XT or 286. However, a 386SX computer doesn't cost much more than a 286 to build. Even if you're just going to use it as a personal home computer, you'll be much happier if you spend a little more and get the best.

Sources

I have been criticized by some for not naming more sources in my first two books. One reason I didn't is because the computer business and technology are so volatile. Every day, a few businesses fall by the wayside; but for every one that fails, a dozen more pop up. I could not possibly name them all, and it's not fair to name one source and possibly overlook another better one.

Another reason that I'm reluctant to name extensive sources is that it might appear that I have some connection with the companies named. I assure you that I have no connection or business affiliation with any company named in this book.

The best source of microcomputer components would probably be a computer store near you that can give you support if something goes wrong. A second choice would be a computer swap, where you can compare and maybe even haggle over prices a bit. A third choice would be mail order. If you look through any of the dozens of computer magazines, you'll find hundreds of good bargains.

I'll have more to say about sources and mail order later in Chapter 15.

Chapter 2

Necessary components

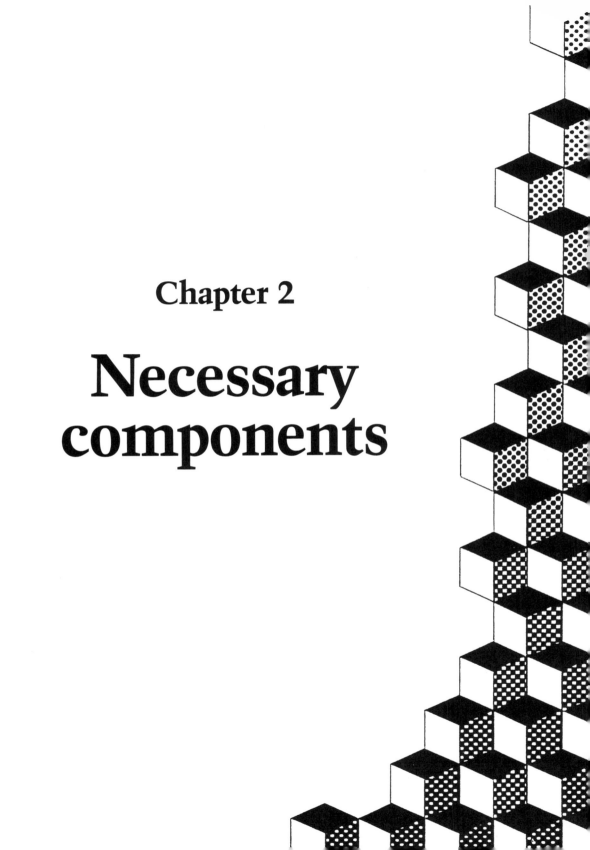

I'm frequently asked how much it costs to build a computer. If you think about it, that's like asking "How much does a car cost?" The cost will depend primarily on what kind of computer you want, what you want the computer to do, whether you assemble it yourself, where you buy it, and how much you want to spend. Choosing from a pool of over six billion dollars worth of hardware on the market today, you can build almost any kind of computer and configure it to do almost anything. (See Fig. 2-1 for some things you might want to put in your 80386.)

2-1 Components needed to build a 386 computer.

Literally, thousands of different options are available to you when you build a computer yourself. Of course, all those choices could be a problem in itself if you have trouble making up your mind as to what you want. I'm going to explain several of the options that'll make it easier for you to decide, although you should be aware that you might never have a completed system. You'll always find new and better things to add to your computer.

Another difficulty in determining the cost of a system lies in the fact that the extensive market competition causes prices to change almost hourly. In this book, I list some approximate costs. However, you can get a better idea by looking at magazine advertisements and checking out the costs of the various components. I've included a list of magazines in Chapter 15 that will help keep you current with this ever-changing technology.

All of the systems—whether PCs, XTs, ATs, or 386a—use similar basic components: they are all plug-compatible. You can even take a power supply or plug-in board out of a true blue IBM (except for PS/2s) and plug it into a compatible, or vice versa, and it will work. The main difference among systems is the motherboard.

386SX and 386DX motherboards

The motherboard is the most important board in the computer: it's the unique board that determines what kind of computer you have. The standard AT-size motherboard has 8 slots, while some of the baby-size boards might have one or two slots less. All of the slots have 62-pin connectors for the standard 8-bit PC bus. In front of some of the 62-pin connectors are additional 36-pin connectors; these connectors are for the 16-bit data lines.

The original 286 and 386 standard motherboard and case was larger than the XT; but with advances in very large scale integration (VLSI), today most motherboards have been shrunk to a size as small as or smaller than the original XT motherboard (see Fig. 2-2).

Many options are available for both the 386SX and 386DX motherboards. The SX motherboards primarily differ in that the 386SX has a 16-bit memory bus while

2-2 A comparison of some motherboards. Left to right, a 386, a 286, and an XT.

the 386DX has a 32-bit memory bus. In addition, the 386SX operates at a frequency of 16MHz, 20MHz, or 25MHz, while the 386DX operates at 25MHz, 33MHz, or 40MHz. (See Fig. 2-3 for a slimline 386SX motherboard.)

2-3 A slim line 386SX with a single vertical daughterboard, with three slots on one side and two on the other.

Both systems might offer options for built-in features and functions, such as on-board cache, IDE hard disk interface, parallel and serial ports, game and mouse ports, and VGA monitor driver. At the present time, the SX motherboards will cost from $250 up to $600.

The ISA 386DX motherboard might cost from $500 up to $900. It might also be configured for the enhanced industry standard architecture (EISA) system, which could cost from $900 up to $1200.

Figure 2-4 shows a VIPC 386 motherboard, while Fig. 2-5 shows a baby 386 motherboard.

Cache

When running a program, the CPU often loops in and out of memory many times, usually using the same memory over and over. If a small cache of high speed memory is set up for the most often-used memory, the system can be sped up considerably. The motherboard must be designed for a cache system, which usually costs a bit more.

Coprocessors

A coprocessor can speed up the processing of certain applications by as much as 500 percent. The software, however, must be designed to take advantage of a

386SX and 386DX motherboards 11

2-4 The VIPC 386 motherboard, my first 386.

2-5 A baby 386 motherboard.

coprocessor. Most of the spreadsheet, graphics, CAD, database, and hundreds of other programs will make use of a coprocessor if one is present. At this time, the cost of a 387SX-16 is about $100, a 387SX-20 is about $120. A 387DX-25 and 387DX-33 cost about $200 apiece.

The motherboard should have a socket near the CPU for installing the coprocessor.

ISA and EISA

The original 8-bit PC and XT boards and motherboard slot connectors had 31 contacts on each side, for a total of 62. This bus was more than sufficient for the power, grounds, refresh cycles, RAM, ROM, and all of the other I/O functions.

When IBM introduced the 16-bit AT, several other new bus functions were added. They needed the 62 pins already in use on the bus, plus several more for the new functions; they also needed a larger connector, but a new connector could make obsolete all of the hardware available at that time. Someone at IBM came up with a brilliant design: they simply added a second 36-pin connector in front of the standard 62 pin. This provided a dual purpose socket for 8-bit boards with 62 or for 16-bit boards with 98 contacts.

When the 32-bit 386 was introduced, they still used the 16-bit AT standard bus; once again, more contacts were needed. Now IBM added a second 62-pin slot connector in front of another one to provide 124 contacts for special 32-bit memory boards. Most manufacturers have abandoned this design. Most boards today have SIMM or SIP sockets with 32-bit bus lines for on-board memory. Except for the 32-bit memory bus, all of the 286, 386, and 486 ISA systems still use a 16-bit bus.

In 1987, IBM introduced their PS/2 line with Micro Channel Architecture (MCA). This system added many new functions to the computer and required more connector contacts. This time, however, they designed a new bus and connector system that was completely incompatible with the billions of dollars worth of available hardware. Users were in about the same boat as the Apple users; they had only one vendor and the prices were very high.

Most people readily admitted that the MCA offered some real advantages over the old original IBM standard. One excellent feature of the MCA system is the Programmed Option Select (POS). The MCA plug-in boards have a unique identification or ID. When a board is plugged in, the bus recognizes it by its ID and automatically configures the board for use with the system interrupts, ports, and other system configurations. In addition, if the board has any switches, it will tell you how they should be set.

MCA offers several other very good features such as bus arbitration—a system that evaluates bus requests and allocates time on a priority level (relieving the CPU of some of its burdens). The MCA bus is much faster than the old AT bus.

MCA offers several other excellent benefits, but IBM is still the single source. A few clones and a few third party MCA boards exist, but they must pay a license fee to IBM and thus are rather expensive.

The Industry Standard Architecture (ISA)

ISA is what used to be known as the IBM AT Standard. For the vast majority of applications, the ISA systems will be more than adequate.

One of the greatest advantages of the ISA systems is the extraordinary versatility and flexibility that it offers. At least 10 billion dollars ($10,000,000,000) worth of IBM-compatible (or ISA) computer components exist.

Most of the ISA components are interchangeable. You can take any board or peripheral from a genuine IBM XT or AT and plug it into any of the clones and it will work; you can also plug any of the components found in a clone into a genuine IBM or another clone and it will work. Because so many clone products and vendors exist, they are readily available almost anywhere and are thus relatively inexpensive (because of the competition).

EISA and the Gang of Nine

IBM's MCA was a system that was no longer compatible with a large number of clone manufacturers. Many large corporate users abandoned the clones and adopted the IBM PS/2 systems because of the added functionality. The cloners were clearly being hurt.

In response, a group of nine compatible makers—soon joined by three others—got together: Advanced Logic Research (ALR), AST Research, Compaq Computer, Epson America, Everex Systems, Hewlett-Packard, Olivetti, Micronics, NEC, Tandy, Wyse Technology, and Zenith. They developed the Extended Industry Standard Architecture (EISA). This new standard includes the bus speed, arbitration, bus mastering, Programmed Option Select, and most of the other functions found on the MCA. Unlike the MCA, however, the EISA bus is downward-compatible. The new EISA standard is designed so that the older style XT and AT boards can still be used; no matter whether your boards are 8, 16, or 32 bit, the EISA bus will accept them.

In some areas, the EISA system will outperform the MCA system. However, one of the biggest advantages of EISA over MCA is that it's compatible with the billions of dollars of earlier hardware.

The EISA connector

We noted earlier that IBM designed the 16-bit AT bus by simply adding an extra 36-pin connector to the 8-bit 62-pin connector. This was an excellent idea because all of the earlier 8-bit boards could still be used.

The AT bus or ISA has 49 contacts per side, or 98 total. Each contact is .06" wide with .04" space between them. The total connector contact area of a board is 5.3" long.

The IBM MCA system needed 116 contacts, but IBM designed the MCA board by miniaturizing the contacts. Each of the MCA board contacts is .03" wide with .02" space between them, which is just half the space between contacts on the ISA boards. They put 116 contacts in a board only 2.8" long, which is about half the size of the ISA 16-bit connector area. Thus, all older ISA boards are completely unusable in the MCA system.

The EISA system also needed more contacts, so they added one hundred more for a total of 198 contacts (which is 82 more contacts than the IBM MCA). Thus, EISA has the opportunity to add more functions than possible with the MCA system. They simply added an extra row of contacts on the plug-in EISA boards by extending the contact area below the standard 98 ISA contacts, connecting the lower EISA contacts with etched lines in the .04" spaces between the 98 ISA contacts.

Unlike the MCA system, the EISA connector is designed so that it is compatible with all of the previous 8-bit and 16-bit boards, as well as the new 32-bit EISA board. They did this by designing a socket that was twice as deep as the original ISA slot or connector socket. This new socket has two sets of contacts, one set at the bottom of the socket for the added contacts and a set at the top portion for the original 98 contacts. At certain locations across the bottom of the socket, there are narrow tab projections that act as keys. These tabs prevent an 8-bit or 16-bit board from being inserted to the full depth of the socket, which allows them to mate with only the upper contacts.

The EISA boards have notches cut in them to coincide with the keys on the floor of the EISA socket. Therefore, the EISA boards can be inserted to the full depth so that both sets of contacts mate in the connector.

Do you need an EISA computer?

At this time, most EISA motherboards are designed for the 486, with only a few 386 vendors. Several more should be around by the time you read this, however.

Your need for an EISA motherboard will depend on whether you will be using it for high end applications such as a network server or a workstation. Because the EISA motherboard provides much more functionality than the ISA, it is considerably more expensive. Still, an EISA system is still much less expensive than an equivalent IBM PS/2 system.

Cost of basic components

Components other than the motherboard would be the same for both the 386SX and 386DX or EISA.

Item	Cost	Item	Cost
Case	$35-105	386SX motherboard	$250-600
Power supply	$40-70	Plus basic components	$650-2920
Monitor	$65-900		
Monitor adapter	$40-200	386SX system total	$900-3520
Multifunction board	$50-200		
Floppy drive, 1.4M	$55-75		
Floppy drive, 1.2M	$55-75		
Hard disk, 80-300M	$250-995		
Disk controller	$20-150	386DX motherboard	$500-1200
Keyboard	$40-150	Plus basic components	$650-2950
Total	$650-2920	386DX system total	$1150-4150

As you can see, the cost can largely vary depending on the particular components. A large variation in cost also exists from dealer to dealer. Some of the high volume dealers charge much less than the smaller ones, so it will pay you to shop around a bit and compare prices. These figures are only rough approximations. The market is so volatile that the prices can change overnight. If you are buying through the mail, you might even call or check out the advertised prices before ordering. Often the advertisements must be written one or two months before the magazine is published, so the prices could have changed considerably.

I have listed several options not absolutely necessary for an operating barebones system. You can buy a minimum 386SX system for less than $900. If you're short of cash or don't need a lot of goodies at this time, you can buy the minimum components and add to it later. For instance, you probably don't need both floppy drives; you can get by fine with only a 1.2M drive. You should also be able to find a 20 or 30M hard drive for about $125. You'll need one much larger later on, but this would get you started.

I will discuss each of the basic components briefly; they will be discussed in more detail in later chapters.

Case

The case usually comes with a chassis and cover, a bag of hardware, a speaker, plastic standoffs, and guides. The case will usually have a switch panel that might or might not be assembled and mounted. The switch panel will usually have a set of keys and a lock that locks out the keyboard without shutting off the power. The panel might have several switches and light emitting diode (LED) indicators with wires that plug into the motherboard and other boards. There will probably be a switch for the Turbo mode, a Reset switch, a Power LED indicator, and a LED for the hard disk activity.

Several different types and sizes of cases are available. There is the original large standard AT desktop type, the smaller XT-size desktop type, and the low-profile desktop type. There are also several floor-standing tower cases: the mini-tower, the medium-size tower, and the full-size tower.

You will only be able to use the low-profile desktop type with a motherboard designed for them. These motherboards have a single slot that accepts a vertical plug-in daughter board. This daughter board has three slots on one side and two on the other side to plug boards in horizontally. (Look back at Fig. 2-3.)

I don't like the low-profile cases; they are about an inch or so lower than the standard height cases but limit you to only five slots. Even then, you might not be able to use the full-length boards. I don't worry too much about the height of my computer case; I have lots of upward space, but I usually don't have too much desk space. The developers of the low-profile systems were probably trying to look as much like IBM as possible. The IBM PS/1 is so small that they didn't have room for the power supply in the case, so they integrated it into the monitor. You can't buy a separate power supply for the PS/1; if a power supply fails in a PS/1, you'll have to buy a $400 monitor to get a power supply.

No matter what type of case you buy, you should make sure that it has at least four bays for mounting disk drives. You will need two bays that are accessible when the cover is installed for the floppy drives. You should have at least two for hard disks. If you expect to mount a tape backup drive or a CD-ROM drive, you will need a couple more bays. Of course, the more bays, the larger the case. The large tower cases may have seven or more bays. Some of the bays might have provision for mounting the drives on edge or vertically. The hard disks can be mounted almost any way except upside down.

Switch panel wires

You might have a bit of trouble connecting the various wires from the switch panel to the motherboard. The case is not usually made by the same manufacturer who made the motherboard, and the wires from the switch panel are not often marked. You'll have to follow the connector back to the LED or switch to determine which one it is. The motherboard will have pins near the front of the board for these connections. The connection for the hard disk activity LED will usually plug into the hard disk controller board or on the motherboard if it has a built-in IDE interface. You should receive some documentation or a diagram with the case that shows where the wires should connect. The motherboard will also have markings near each pin connector.

Power supply

Many vendors sell the case and power supply as a single unit. Most units sold today have a capacity of at least 200 watts, which should be sufficient for most ordinary systems.

The early power supplies were rather large. If you opened one up, you would find a lot of empty space in the case. Many vendors are now offering a much smaller supply with the same 200-watt rating.

Uninterruptible Power Supply

The power supply takes the 110 volts of alternating current from the wall socket and transforms it down to plus and minus 12 volts of direct current and plus and minus 5 volts of direct current. These are the voltages needed to drive the computer.

We don't usually have thunder and lightning in San Francisco and Los Angeles, so I have never worried too much about an uninterruptible power supply. Just recently, however, I was working on one of the chapters of this book and a freak electrical storm came up. I had worked a couple of hours, polishing and redoing a whole chapter. Suddenly a large clap of thunder exploded directly over my house. Although the lights dimmed for just a fraction of a second, my computer screen went blank. All of the work that I had done in the last two hours was gone.

I don't have to tell you that I was just a bit unhappy, but mostly with myself: I shouldn't have been working with my computer during a storm. Still, my deadline was rapidly approaching. Because I had to work, I should have followed the advice that I have preached for years: BACKUP, BACKUP, BACKUP! Backing up is so easy to do. With WordStar, all I have to do is press the F9 key to save any work on the screen or in memory to the hard disk. Had I taken the fraction of a second that it takes to press F9, I could have saved hours of work.

If backups still seem inconvenient to you, you can also buy an uninterruptible power supply (UPS). Without a UPS, if there is a brief interruption of the 110 volt power—even for just a fraction of a second—any data in the computer's memory is lost. A UPS can take over during an interruption and continue to feed power to the computer until you can save your data to disk or complete what ever operation you are doing. Different types of UPS exist, varying in how much wattage they can supply and for how long. Some might be able to supply enough current for two or three computers for as long as two hours. Others may supply a single computer for 5 to 20 minutes only. Of course, they also have different prices that range from $150 to over $5000. If you live in an area where electrical storms or power outages are frequent and your work is critical, you might consider buying one.

Two types of systems exist: standby and on-line. The *standby* system uses a 12-volt battery and battery charger. The 12 volts is fed into an inverter that converts the 12 volts dc from the battery back to 110 volts ac. The standby usually has a sensor that can detect a power loss, and it immediately switches to supply 110 volts to the computer. There can be a delay of a few milliseconds before the standby can switch over.

The *on-line* types are in series with the 110 volts and provide power to the computer at all times. The 110 volts feed a charger that keeps a 12-volt battery charged. The battery converts its 12 volts back into 110 volts and continuously feeds it into the computer. The advantage is that there is no delay in the event of a power outage. The 12-volt battery will continue to supply 110 volts to the computer until it becomes discharged.

Some companies are now integrating a small rechargeable battery inside the power supply; if the power fails, it can take over and run the computer until the contents of the RAM is saved to disk. The Emerson Company's AccuCard system uses a small card with a rechargeable battery that plugs into one of the eight slots. This card can take over in the event of a power failure and provide power long enough to save the data to disk.

Here is a short list of vendors and their telephone numbers. If you need a UPS, call them for their specifications and latest prices.

Clary Corporation	(818) 287-6111
Controlled Power	(800) 521-4792
Emerson AccuCard	(800) 222-5877
Electronic Specialists	(800) 225-4876

Elgar Corporation   (619) 450-0085
Kalglo Electronics   (800) 524-0400
Para Systems   (800) 238-7272
Perma Power   (800) 323-4255
Sola   (312) 253-1191

RAM memory sockets

Most of the 386SX motherboards have provision for 8-16M of memory on board. They will have sockets for one or more of the three types of memory: dual in-line package (DIP), single in-line package, or single memory modules (SIMM). This memory communicates with the CPU over 16-bit bus lines.

Most of the 386DX motherboards will have sockets to provide up to 32M of on-board memory. This on-board memory communicates with the CPU over 32-bit bus lines.

You can use the older existing 8-bit type plug-in boards that were developed for the PC, XT, and AT with the 62-pin edge connectors. They can be plugged into any of the 62-pin slots in the 386. The same is true for the 16-bit boards developed for the AT.

Monitor

You can buy a monochrome monitor for $65. They are perfectly good for ordinary usage, but I don't know why anyone would put a small monochrome monitor on a fantastic 386. I prefer color, even for word processing, and I also prefer high resolution. If necessary, I would use the rent money and maybe skip a few meals in order to have color.

You can get a fairly good VGA monitor for about $250. The better brand name ones will cost from $500 on up to several thousand, depending on size and resolution. I'll go into more detail on monitors in Chapter 8.

Monitor adapter

The monitor must have a plug-in board to drive it. Several boards are on the market, costing from $40 up to over $400. Adapters are covered in more detail in Chapter 8.

Multifunction boards

If your motherboard does not have built-in functions, you will need a board to drive your printer. You will also need ports for a modem, an extra printer, or a mouse. Several different kinds of multifunction boards have been made. Most of them have functions such as serial and parallel port, while some might have an IDE interface, floppy disk controller, VGA, and several other features. Depending on the functions and utilities that are on the boards, they will cost from $50 up to $200.

Floppy drives

Many vendors are still selling 360K and 720K floppy disk drives, but these drives are obsolete. The 1.2M (5.25") will read and write to the 360K as well as 1.2M, and the 1.44M (3.5") will read and write to the 720K as well as the 1.44M. If you only buy one drive, make sure it is a 1.2M.

Hard disks

You can run a 386 on floppies alone without a hard disk, but I can't imagine why you'd want to do that. The hard drive is one of the essential elements that helps make the 386 so powerful.

Several hard disk vendors are in business, selling many types and grades of quality. Hard disks are such an important part of the computer system that I have devoted all of Chapter 6 to this subject.

Basically, you should select a well-known brand with an access speed as fast as possible and, most important, with the largest capacity that you can afford. At one time, 10M was a large disk. Nowadays, that isn't enough to store more than one or two software packages; there seems to be an immutable law that says that data and files will multiply and increase on a hard disk until it is completely filled.

Disk controllers

The disk drives, both hard and floppy, must have a controller. The better controllers have both hard and floppy controllers on a single plug-in card. I'll say more about controllers in Chapters 5 and 6.

Keyboards

Data can be input to a computer in several different ways such as by modem, by scanners, by a mouse, and by voice. Still, the keyboard is by far the most often used and most important input device. It is your most intimate contact with your computer.

Several different types of keyboards are on the market. Some might cost as little as $50, while others range up to $250. In my opinion, some of the low-priced ones seem to do as well as the high-priced ones.

If at all possible, try out the keyboard before buying it. Some keyboards have soft and sensitive keys. I have a heavy hand and thus have great difficulty using these types. Some of the better models have different springs that you can order and install under the keys to change them to suit your style. I'll have more to say about keyboards and other input devices in Chapter 10.

Where to find the parts

Most large cities have computer stores. Check the advertisements in the daily and Sunday newspapers and in the telephone directory. Keep in mind that most retail computer stores usually sell items at or near the suggested list price, which could be nearly twice as much as the "street" price. You must shop wisely.

If you live near a large city such as Los Angeles, San Francisco, New York, or Dallas, you probably know of a computer swap going on almost every weekend. Swaps are one of the better places to shop. You can compare prices, look the parts over, and even try them out in some cases. You could even possibly haggle a bit with the dealers, especially near closing time. Some will even sell the items at or near their cost rather than pack them up and cart them back to their main store.

Alternately, you could look through the mail order advertisements in the computer magazines. I've heard some horror stories about a few unethical mail-order companies in the past. Still, mail-order and computer advertisements are the life's blood for computer magazines; a few phony ads can ruin their reputations. The magazines today are very careful about whom they advertise. In my opinion, the vast majority of the advertisers are honest and will deliver as promised.

I'll explain more about mail order in Chapter 15.

Chapter 3
386SX or 386DX assembly

Figure 3-1 shows my first 386 motherboard. I bought it from a friend; but even with a good discount, it still cost me $1825. It was one of the better designed motherboards at that time, however; it had several built-in features, 1M of memory, and a standard 16MHz 386DX CPU that had been selected to run at 20MHz. Still, the prices have been dropping; ironically enough, I recently bought a 20MHz 386SX motherboard for only $250.

3-1 The VIPC motherboard.

Benchtop assembly

When I first assemble a computer, I connect everything together on a bench top or the kitchen table before installing it in the case. It only takes me a few minutes to connect the power supply to the motherboard and the disk drives. Then I connect the drives to the controller and connect the monitor. Finally, I turn the computer on and make sure that everything works. If any problems occur, I can find them fairly easily. Everything usually works, though, so I then install it in the case.

Assembly instructions

Figure 3-2 shows the pins for the built-in on-board parallel and serial ports and the cables. The cable being installed is a serial port for COM2, and the pins for COM1 are directly in front of it. The parallel port—LPT2—has a cable installed to the left, and LPT1 is directly in front of it. The pins to the right of COM2 are for the built-in EGA monitor driver output. To the right of that are the pins for the light pen.

3-2   Installing port cables.

Figure 3-3 shows the battery pack being connected; it plugs into pins near the keyboard connector. The battery holder is attached to the power supply by a 2" square of double-sided foam tape. This battery pack is just a backup for the on-board rechargeable battery. The newer systems no longer offer this option; most systems today have a small tubular rechargeable battery soldered to the motherboard.

3-3 Installing the battery cable.

Figure 3-4 shows the main power to the board being plugged in. You should be careful here because you could plug these two connectors in backwards (not a good idea). When installed properly, the four black wires are in the center.

Figure 3-5 shows my helpmate installing the hard disk controller. Figure 3-6 shows the controller cable for the hard disk being connected to the controller board. Be very careful because this connector can be plugged in backwards, possibly destroying the electronics on the hard disk and the controller board. This 34-wire ribbon cable has a colored wire on one side that indicates pin one on the connector. Also, this cable might have a connector in the middle of it (for a second hard disk drive, if one is present).

Figure 3-7 shows the 20-wire data cable being connected. Note that it too can also be plugged in backwards. It has a colored wire on one edge that indicates pin one. If only one hard disk is used, the row of pins nearest the 34-wire connector is used. If a second hard disk is installed, then there will be a second 20-wire cable to connect to the row of pins near the front of the controller board.

Your controller board might be different. You should have received some sort of documentation with your board; check it and follow the instructions.

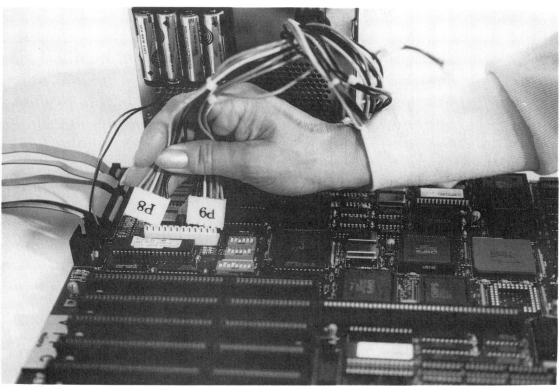

3-4 Installing connectors for motherboard power. Note that four black wires are in the center when properly connected.

3-5 Installing the hard disk controller card.

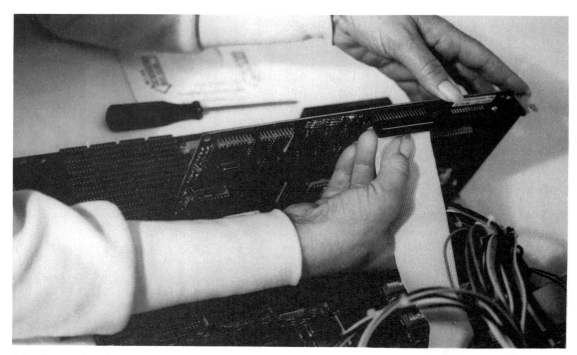

3-6 Connecting the 34-wire ribbon cable to controller card. Make sure that the colored wire on the edge of cable goes to pin one on the board.

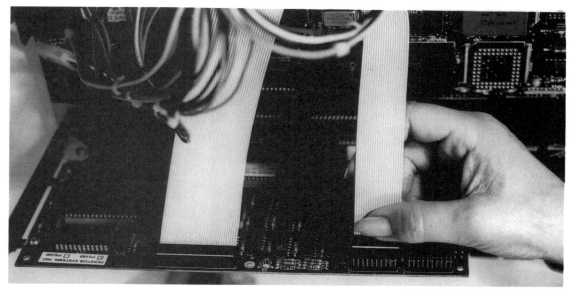

3-7 Connecting the hard disk data cable to the controller card. Make sure that the colored wire goes to pin one.

Figure 3-8 shows a 34-wire ribbon cable being connected to the floppy disk controller. Again, you must be careful because it can be plugged in backwards as well. It should have a different colored wire (indicating pin one) that should go to pin one on the board.

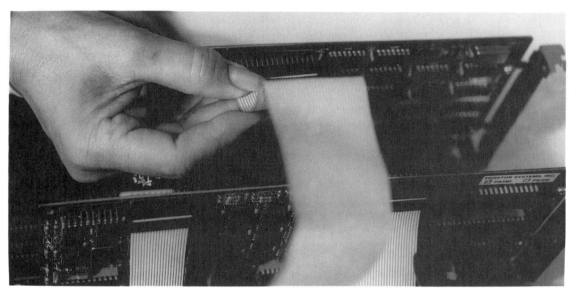

3-8 Connecting the floppy disk controller cable. Make sure that the colored wire goes to pin one.

Figure 3-9 shows the motherboard with the controller boards and cables installed.

Figure 3-10 shows the plastic slides that must be installed on the disk drives. The several holes in the slide allow it to fit almost any case. You should put in a single screw to hold one slide, and then try it in the case to make sure you have the right holes before installing all of them. Figure 3-11 shows the installation of the slides. Many of the newer cases do not have provisions for the slides shown here; most of the drives are now mounted directly in the bays with screws.

Figure 3-12 shows the connection of the controller cable to the hard disk. This is a Seagate MFM hard disk. Note that it has both a 34-pin controller connector and a 20-pin data connector. The IDE drives have a single 40-pin connector. Look for the colored wire and an indication of pin one on the hard disk edge connector. Some cable connectors are keyed so that they can only be plugged in properly. The key consists of a bar installed in the face of the connector so that it slides into the cut slot in the edge connector. The edge connectors for both floppy and hard disks have a keying slot cut between pins 2 and 3. If the connector does not have a key, you could plug it in backwards (still not a good idea). Look for the slot between pins 2 and 3; this is a good indication of which side pin one will be on.

28 *386SX or 386DX assembly*

3-9 All controller cables are connected to the boards.

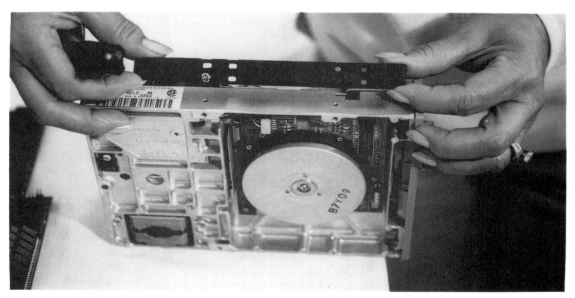

3-10 Preparing to attach the plastic slide rails to the disk drives.

3-11 Attaching the slide rails. Install one rail and then temporarily place it in a case slot to make sure that it is in the right holes before installing all of them.

3-12 Connecting the controller cable to the hard disk. The colored wire should go to pin one.

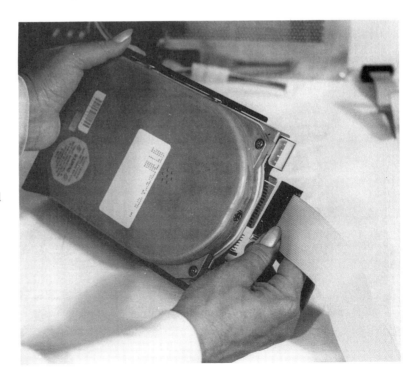

30 386SX or 386DX assembly

Figure 3-13 shows the connection of the 20-wire data cable to the hard disk. Again, it can accidentally be plugged in backwards. Take care that the colored wire goes to pin one on the edge connector. This edge connector will also have a keying slot between pins 2 and 3.

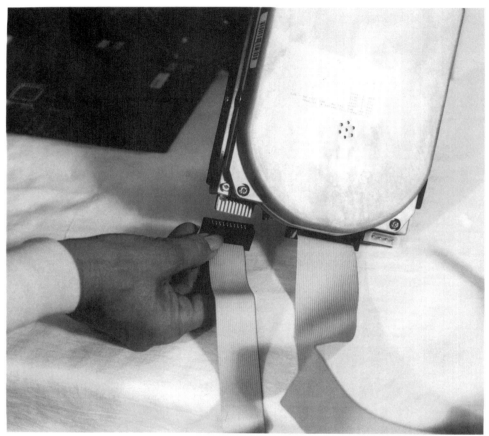

3-13 Connecting the data cable to the hard drive. The colored wire should go to pin one.

Figure 3-14 shows the connection of power to the hard disk. The power supply has four cables for disk power. They are all identical and can only be plugged in properly.

The small 3.5" floppies require a small adapter for the power. Figure 3-15 shows this adapter being plugged into the standard power cable.

Figure 3-16 shows the connection of the small adapter power cable to the 3.5" floppy. It can only be connected properly.

Figure 3-17 shows the middle connector being attached to the 3.5" drive. It can accidentally be plugged in backwards. Match the colored wire with pin one on the edge connector. Note that this edge connector also has a keying slot between pins 2 and 3.

Benchtop assembly 31

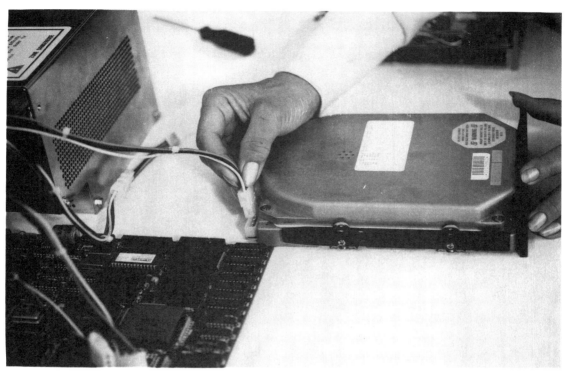

3-14 Connecting the power cable to the hard disk.

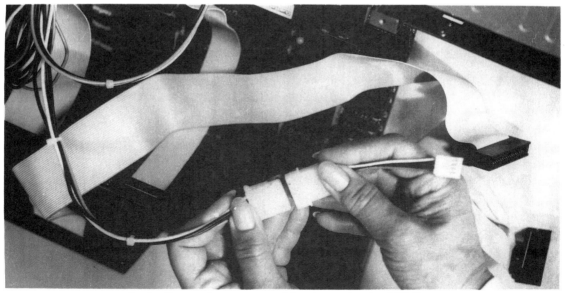

3-15 Connecting a power cable adapter for the 3.5″ floppy.

386SX or 386DX assembly

3-16 Connecting the 3.5″ power cable to the drive.

3-17 Connecting the middle connector to the 3.5″ drive. The colored wire goes to pin one.

The floppy controller cable has three connectors, one on each end and one in the middle. Because the 3.5″ floppy will be my B drive, it attaches to the connector in the middle.

Figure 3-18 shows the connection to the 1.2M floppy drive. This will be my A drive, which is indicated by the split and twist in the cable. Once again, this connector can also be plugged in backwards. Look for pin one, or the keying slot between pins 2 and 3, on the edge connector and make sure that the colored lead of the cable goes to that side.

3-18 Connecting the floppy controller cable to drive A. This is the end connector with the split and twist in the cables, which indicates drive A. The colored wire goes to pin one.

Figure 3-19 shows the power connection to the A drive, and Fig. 3-20 shows all drive cables connected. Figure 3-21 shows the connection for the monitor. Figure 3-22 shows the keyboard connection at the rear of the motherboard. Figure 3-23 shows the assembled system, ready to start.

If you don't already have DOS, you will need to buy a copy. You can use either DR DOS 6.0 or MS-DOS 5.0. If you buy the MS-DOS 5.0, make sure you don't get the upgrade version—it doesn't have a boot disk. (You can make a boot disk after it

3-19 Connecting a power cable to drive A.

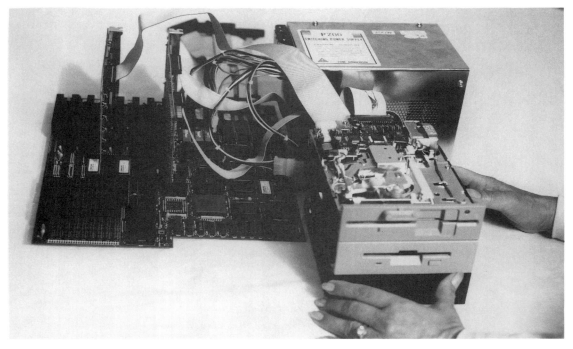

3-20 All cables connected to the drives.

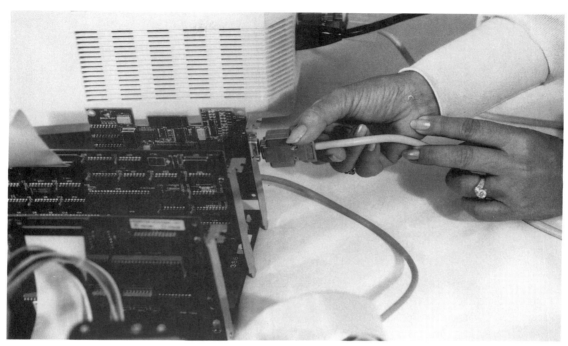

3-21 Connecting the monitor cable.

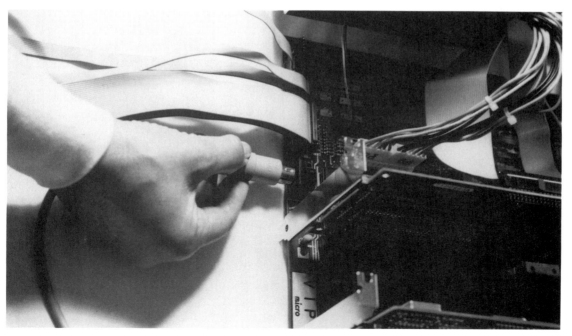

3-22 Connecting the keyboard cable.

36 386SX or 386DX assembly

3-23 All components connected and simply waiting for power.

has been installed on the hard disk; but if you can't boot up, you can't make a boot disk.) In order to boot the computer, you must have a bootable copy of DOS on a diskette for the A drive. You can make either the 5.25″ or 3.5″ drive the A drive by attaching the end connector with the twist to that drive. If you have both drives, it is usually best to make the 5.25″ drive the A drive.

Installing the components in the case

After running the system on the bench with no problems, I installed it in the case. Figure 3-24 shows us preparing to mount the power supply. Note that there are two raised tongues in the floor of the case and two matching cutouts on the bottom of the power supply case. The power supply is placed over the raised tongues and then slid towards the back of the case. Figure 3-25 shows me installing four screws in the back of the case, which secures the power supply.

Figure 3-26 shows the back side of the motherboard as I'm preparing to slide it into place. The four white objects are plastic standoffs that slide into grooves in the raised channels on the floor of the case.

In Fig. 3-27, I have just run into an installation problem. There are three black plastic standoffs with a horizontal groove that accepts the right edge of the motherboard. As you can see from the photo, one of the standoffs is exactly where the light pen connector on the motherboard is located. This is a slight design fault of the board; I need support for this edge of the board. I could drill a new hole and

Installing the components in the case 37

3-24 Installing the power supply in the case. Note the raised tongues on the floor of the chassis and the matching cutouts on the bottom of the power supply; these tongues hold the power supply down.

3-25 Two screws in back of the chassis hold the power supply in place.

38 *386SX or 386DX assembly*

3-26 Preparing to slide the motherboard into the case. The white objects are plastic standoffs that slide into grooves on raised channels in the case.

3-27 A minor problem. A black plastic standoff is interfering with the connecting pins for the light pen.

relocate the black standoff, but I found an easier solution instead. As shown in Fig. 3-28, I solved the problem by using a pair of cutters and cutting away some of the plastic standoff. The board now slips in easily.

Figure 3-29 shows me installing one of the two screws used to hold the board in place. The other screw is at the front of the board. Note that the power connection has the black wires in the center.

3-28 Solution to the problem. I used a pair of sharp snips to cut away part of the top of the standoff.

3-29 The motherboard is installed and power is connected. A screw is being installed at the rear center of the board. One other screw will be installed in the front of the board to hold it down.

Switch panel wires

Also note that a lot of loose wires are at the front of the case. These wires are from the switches and LEDs on the front panel that came with the case. In some instances, there might be 15 or 20 wires from the switches and LED indicators. You will probably have to trace the individual wires back to the switch or LED that it comes from, and then find the pins on the motherboard for that connection.

Your motherboard and system might not use all of these wires and switches. You should get a manual with your motherboard that indicates where each of the wires should be connected. Most hard drive controllers have a connection for the hard drive activity.

Figure 3-30 shows all components installed, ready for the cover. In Fig. 3-31, the system is up and running.

3-30 All components are installed in the case, ready for the cover.

TEST.BAT

Once I have installed it in the case, I turn the system on and let it run for a couple of days to burn it in. A semiconductor has no mechanical moving parts. If a circuit is designed properly and the semiconductors are manufactured properly, then a semiconductor should last several lifetimes. If a semiconductor is going to fail, it will usually do so in the first few hours of use. This is called *infant mortality*. A good burn-in will usually weed out the chips that are subject to infant mortality.

Once everything was working properly, I made a short batch file that would help exercise the system for burn-in. Here is the short file that I typed at the C prompt:

3-31 Up and running.

```
COPY CON TEST.BAT
DIR
TEST2 .BAT
^Z
```

When I finished, I pressed F6 to end the batch file. (Ctrl-Z will do the same thing.)

I then typed this at the C prompt:

```
COPY CON TEST2.BAT
TEST
^Z
```

This sets up a loop. It runs the DIR command (which displays the directory) and then calls TEST2.BAT, which calls the TEST.BAT. This loop will run continuously and can be stopped by pressing Ctrl-C. There are more elegant and sophisticated ways of testing, but this simple test does a fairly good job.

Slot covers

You might not use all of the available slots, but the openings at the back of the case should be covered. Usually the only noise you hear from your computer is the fan in the power supply, which is supposed to draw cool air in from the open grill in the front of the case, pull it over all the boards and components, and then exhaust it out of the opening at the back of the power supply. If several other openings are at the front of the computer, air will be drawn from them and might not accomplish the necessary cooling.

Heat is an enemy to semiconductors. Don't put anything in front of the computer that would cut off the circulation or in back of it that would prevent the outflow of air. Your computer will last a lot longer if you keep it from losing its cool.

Chapter 4

Upgrading your computer

If you are stuck with an old PC, XT, or 286, you might be perfectly happy if it does all that you need it to do. However, you might feel like a person putt-putting around in a Model T when everyone else is zipping around in new Cadillacs. You might have been dreaming of the luxury of the new 386s, so you might want to get rid of your old clunker. The trouble is that there isn't too much of a market for used PCs. If your machine has an IBM logo on it, you might have paid from $3000 up to $5000 for it a few years ago. Today, though, you might be lucky to get $100 for it today. You could donate it to a charity, but the IRS probably won't let you deduct as much as you think it's worth from next year's taxes.

Baby-size motherboards

You can easily salvage some of your investment. For about $250, you can install a 386SX motherboard in the old case; for about $400, you can install a genuine baby 386DX in it. It's easy to do, only takes about 20 minutes, and will add up to $1500 to its value.

Shortly after IBM released their 80286 AT in 1984, the Chips and Technology Company began designing Very Large Scale Integration (VLSI) chips that could integrate several of the AT motherboard functions. They were able to design a single chip that could integrate and replace as many as 30 chips on the standard-sized AT motherboard. Using these chips, the standard-sized AT motherboard was reduced to the size of the XT motherboard; these smaller motherboards were called *baby-sized*.

Several other companies besides Chips and Technology now also produce VLSI chipsets. Almost all motherboards are now designed to use them, including the 486. These motherboards are functionally equivalent to the standard-sized boards but are a bit more reliable than the standard size because so many functions are integrated into a single chip. The more connections inside a chip, the less traces and solder connections exist on the board. You can use one of these boards to replace the original motherboard in an old PC, XT, or 286, and turn it into a more powerful 386SX or 386DX machine.

Cost to convert

The motherboard will be one of the most expensive components in your new 386 machine. You can buy XT motherboards for less than $50; but, depending on what you choose, the 386SX motherboard will cost between $250 and $500. The 386DX will cost between $400 and $1200. You might have some problems with the case, power supply, keyboard, and old memory, but you should be able to use your disk drives, plug-in boards, and other peripherals.

Case

If you are converting an old IBM original PC, you might have some case alignment problems. The original IBM PC has only five slots, while the XTs, 286s, and 386s usually have eight slots. Thus, the openings on the back of the PC case with only five slots will not line up with the eight slots of the 386.

You could buy a new case for about $30. You should have no problems installing a baby 386 motherboard in an XT case. Neither should you have problems installing a baby 386 in a standard-size AT or 286. You will have a lot of empty space, but the slots should line up with the rear panel openings.

The standard PC or XT case size is 19.6" wide by 16" deep by 5.5" high. The standard AT case used for the 286 and 386 is 21" wide by 16" deep by 6.25" high. Several manufacturers are now also making special cases with the same footprint as the XT but an inch higher to accommodate some of the larger 16- and 32-bit boards.

Power supply

One other problem that you might have in converting a PC or XT into a 386 is the power supply. Most of the original IBM PCs were sold with a 65-watt power supply. The XTs were sold with 135- or 150-watt supplies. For a 386, you will need to upgrade to about 200 watts, which will cost about $50.

Keyboard

Still another problem will be the keyboard. The PC and XT keyboards won't work with a 286 or 386. Both keyboards have the same type of connector; however, the keyboard itself is a small computer, and the scan frequency of the PCs and XTs is different than the 286s and 386s. Some of the clone keyboards have a switch that allows them to be used with both types. You can buy a good keyboard for as little as $45.

Here is a sample cost for a conversion. The total price for such a conversion will vary a great deal, of course, depending on what you already have and what you choose to put in your new computer.

Of course, you might want to install new floppies, a new hard disk, a new monitor, and other goodies. You can always add these components later if you don't have the money at the moment.

Item	Cost
Baby 386 motherboard	$250-1200
Power supply	$50-70
Keyboard	$45-100
Total	$345-1370

The conversion of an IBM XT

Figure 4-1 shows me stripping down an IBM XT, while Fig. 4-2 shows the case with the motherboard removed. I left the original drives installed in the case

46 *Upgrading your computer*

4-1 Stripping down an IBM XT in order to install a baby 386 motherboard.

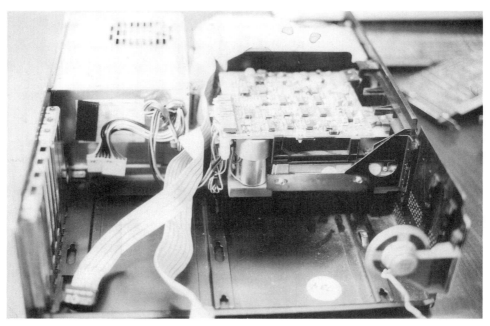

4-2 Ready to install a new motherboard. Note that you don't have to remove the disk drives and several other items.

because I knew that these drives were good. I will use some other drives to bench-test our motherboard. Note that this is a fairly late model XT, as evidenced by the raised channels that have slots for the plastic standoffs of the motherboard.

Figure 4-3 is my baby 386DX motherboard. Note that the board has eight large Chips and Technologies VLSI chips. The large chip in the lower left corner is the 80386 CPU. Notice that it has 8M of 32-bit memory in the Single In-line Memory Modules (SIMMs) installed on board in the upper-left-hand corner.

Before I install the motherboard, I will connect all the components on the bench to make sure it works. The instructions for the cable connections will be the same as those outlined in Chapter 3. They are important, however, so I will repeat them here.

4-3 A baby 386DX motherboard.

The first thing to do is to connect the power supply as shown in Fig. 4-4. Note in Fig. 4-5 that the four black leads are in the center. Figure 4-6 shows a monitor adaptor board and a combination floppy and hard disk controller with cables installed on the board. All of the cables have a colored wire on one edge, which indicates pin one. The floppy cable is plugged into the controller board on the row of pins nearest the back. Pin one is at the top of the board, so the connector is plugged in so that the colored wire is on top.

Figure 4-7 shows a 3.5″ floppy being connected as drive B to the middle connector. The edge connector on the floppy disk board should have a 1 etched on one side and a 2 on the other side of the board. Alternately, look for the keying slot between pins 2 and 3. Make sure that the colored wire goes to this end. Figure 4-8 shows the power cable being connected to the 3.5″ floppy. It can only be plugged in one way.

48 *Upgrading your computer*

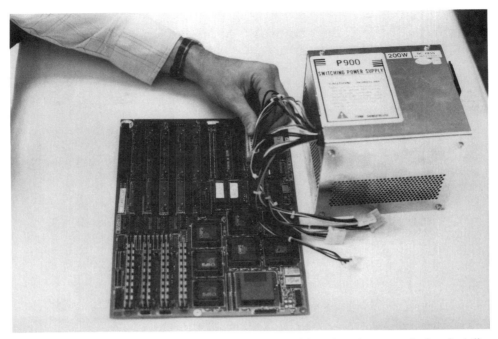

4-4 Connecting the power supply to the motherboard for a bench-top test before installation in the case.

4-5 Showing the correct installation of the power cables with the four black wires in the center.

The conversion of an IBM XT **49**

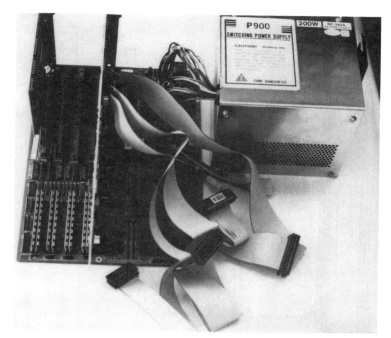

4-6 A combination hard and floppy drive controller card with all cables connected. On most boards, the floppy cable will be connected to the pins nearest the rear of the board.

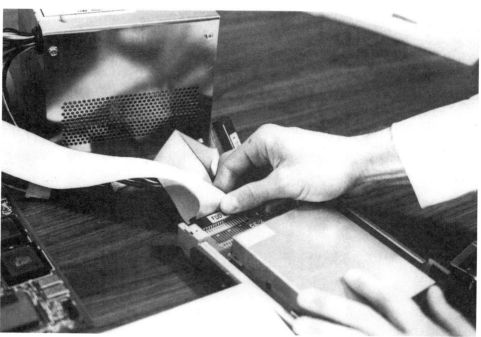

4-7 Connecting a 3.5" 1.44M drive to the board. This drive will be drive B, so it attaches to the connector in the middle of the cable. Make sure that the colored wire on the cable goes to pin one.

50 *Upgrading your computer*

4-8 Connecting the power cable to floppy drive B.

Figure 4-9 shows the ribbon cable for floppy drive A being connected. I know this connector goes to the A drive because some of the wires in the cable have been twisted at this end connector. Again, make sure that the colored wire goes to the side of the edge connector marked 1 or 2 or with the keying slot. Figure 4-10 shows the connection of the power cable to the A drive.

Figure 4-11 shows the connections to a hard disk. It will have three cables, a 34-wire ribbon cable, a 20-wire cable, and the power cable. The 34-wire cable also has a colored wire that indicates pin one or the keying slot between pins 2 and 3. Look for a marking of pin 1 on the hard disk controller board. If you are adding a second hard disk drive, connect it to the connector in the middle of the cable. The connector for the hard disk might or might not have a twist on the end of the cable. If it does have a twist, it will look very much like the cable for the floppy drives. However, different wires are twisted on the hard drive cables, so they are not interchangeable.

Figure 4-12 shows the 20-wire ribbon data cable with the colored wire on the pin one side. Look for a 1 or 2, or the keying slot, on the edge connector and plug the cable connector in to match it. If there is a second hard disk to be installed, it will also have a separate 20-wire ribbon cable. There will be two sets of pins on the controller for the 20-wire data cables. If they are not marked, the set of pins closest to the 34-wire connection is usually for the first hard disk.

Note that this hard disk requires a 34-wire controller cable and a 20-wire data cable. The IDE hard drives have a single 40-wire cable.

4-9 Connecting a 1.2M floppy as drive A. Note that the cable has the split and twist on this end connector. The colored lead should go to pin one.

4-10 Connecting the power cable to the 1.2M floppy.

52 *Upgrading your computer*

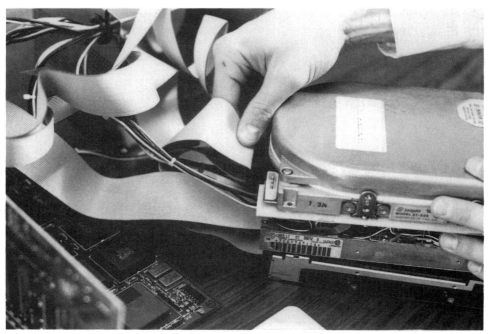

4-11 Connecting the controller cable to the hard disk. The colored wire goes to pin one.

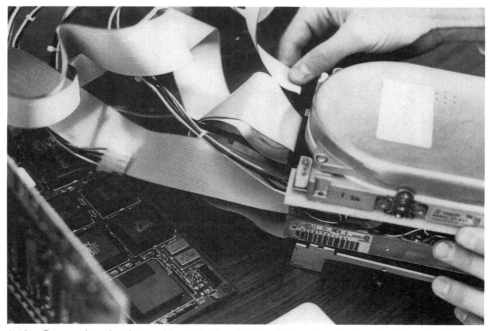

4-12 Connecting the data cable to the hard drive. The colored wire should go to pin one.

The conversion of an IBM XT **53**

Figure 4-13 shows power connection to the hard disk. Figure 4-14 shows all the components connected, ready for our initial bench test.

The completed system worked like a charm, so I installed it back in the case. The owner had a good 20M hard disk and an original IBM full-height 360K floppy drive, so he decided to continue to use these drives for the present time.

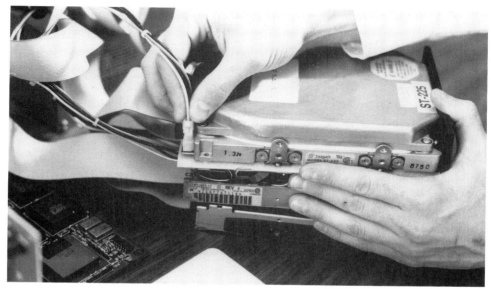

4-13 Connecting the power cable to the hard drive.

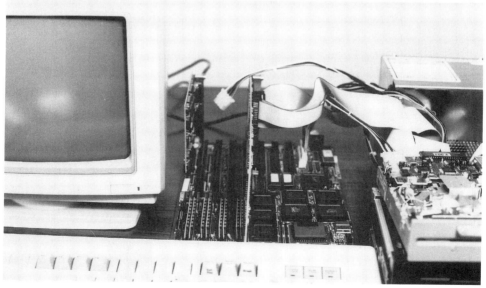

4-14 All components are connected and ready for burn-in.

The IBM logo is still in place on that machine; although the system is no longer an IBM, the owner wouldn't give up the logo. He works in a large office containing several other IBM XTs, and he didn't tell anyone about the update. Thus, on the outside, the microcomputer looked like all of the others, just a mild-mannered XT (like Clark Kent); on the inside, however, it was Supercomputer. Often people would come by to use it and be shocked by its power. He got an enormous kick out of watching their reactions when they first tried it, saying that the reactions alone were almost worth the price of the update.

Accelerator boards

Several companies make plug-in boards that can essentially turn a PC, XT, or 286 into a 386 machine. Adding an accelerator board is usually very simple: just plug in the board and replace the CPU. You will be able to process large spreadsheets, CAD programs, and other CPU intensive programs at the speed of a standard 386. Many of these boards have provision for installing 32-bit memory on board but might be limited as to the amount that can be installed. (See Fig. 4-15.)

4-15 A Quad 386XT accelerator board. This is a fairly inexpensive way for you to experience many of the benefits of the 386.

A drawback to the accelerator board approach is that the XT motherboard has only an 8-bit bus. You will not be able to use any of the plug-in boards designed for the 16-bit 286 and the 386. Another consideration is that the accelerator boards usually cost as much or more than a 386SX or a 386DX motherboard.

It would take only a few minutes longer to install a 386SX or 386DX motherboard, which would give you much more utility and functionality. I do not recommend the accelerator boards.

CPU daughter boards

Some 32-bit system motherboards now being manufactured allow five different options for easy upgrades. A 386DX motherboard could start out with a relatively low cost 25MHz CPU daughter board that could later be replaced with a 33MHz, 40MHZ, 486SX, or 486DX board.

No similar system is made for the 386SX because of its 16-bit bus.

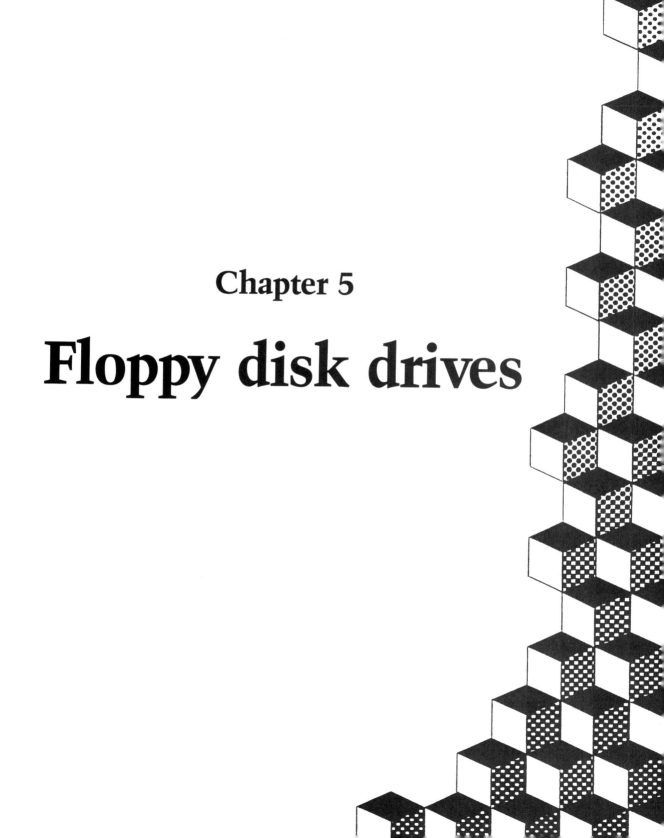

Chapter 5
Floppy disk drives

Floppy disk drives are among the most important components of your system. You have probably heard the terms *PC-DOS* and *MS-DOS* many times. DOS is an acronym for Disk Operating System, with the D representing the floppy disk (which was the only storage available for early systems). The primitive hard disks at that time were only available on very expensive mainframes. PC is a term adopted by IBM for Personal Computer, while MS are the initials of Microsoft Corporation, which developed the disk operating system for IBM. IBM dropped the D from their OS/2 (for Operating System/2 software).

It's been over 10 years since Bill Gates developed MS-DOS 1.0. We are now past MS-DOS 5.0. Bill still looks like a teenager and still has all of his hair, but he's now the head of one of the largest corporations in the world and a billionaire. Bill Gates and DOS have come a long way.

Before Bill Gates and IBM created the DOS standard, most personal computers used the CP/M operating system. CP/M was an acronym for Computer Program/Microcomputer and was developed by Gary Kildahl, now the head of Digital Research. The story is that IBM tried to contact Gary to develop DOS but he was away from home; thus, fame and fortune went to Bill Gates.

At that time, about 100 different personal computers (such as Apple, Osborne, Kaypro, and Morrow) were on the market. Most of them used CP/M, but each one used a different format for their floppy disks. You could not record a disk on one machine and read it on another; they were all incompatible with each other. Software vendors had to provide different copies of the available programs for each machine format. The industry was in a condition similar to that of the Tower of Babel.

MS-DOS and IBM created a standard that almost all manufacturers adopted. You could now record on one machine and read it back on another. This one factor did more to create the computer revolution than any other.

Many of the commands in the early DOS system were quite similar to the CP/M system. Had Gary been as litigious as Lotus concerning the "look-and-feel" of similar software, Gary might now be a billionaire. Still Gary (and Digital Research) are not doing too badly. Digital Research has cloned MS-DOS, releasing DR DOS 5.0 almost a year before Microsoft released their MS-DOS 5.0. Some experts say that DR DOS 5.0 is superior to MS-DOS 5.0 in several ways, and Digital Research has now released DR DOS 6.0.

Some floppy disk basics

Floppy disks are made from a plastic material called polyethylene terephthalate. It is coated on each side with a thin layer of magnetic material made primarily from iron oxide (in layman's terms, powdered rust). Bits of cobalt and other materials are added to give it special magnetic characteristics. The finished product is similar to the tape used in audio cassette and video tape recorders.

The following concepts are generally used with floppy drives and disks.

Tracks

A disk is similar in some respects to a phonograph record. A record has only one track that starts at the outer edge and winds toward the center, but a 360K floppy has 40 single concentric tracks on each side. The first track is track 0 and the last is 39. The 1.2M and 3.5" disks have 80 concentric tracks on each side, numbered from 0 to 79.

Formatting

When you buy a floppy disk, you must format it before it can be used; it's blank when it comes from the factory. Formatting records data on the disk to mark and number each track and sector.

Sectors

Each of the 40 tracks per side of the 360K and the 80 tracks per side of the 720K are formatted so that each track is divided into 9 sectors. The 80 tracks per side of the 1.2M high density floppy is divided into 15 sectors, and each of the 80 tracks per side of the 1.44M is divided into 18 sectors. Each sector holds 512 bytes.

If you do the math for a 360K disk, it would be 40 tracks × 9 sectors × 512 bytes × 2 sides = 368,460 bytes. During formatting, 6144 bytes are used to mark each track and the beginning of each sector. Thus, 368,460 − 6144 = 362,496 bytes of usable space (normally rounded to 360,000 or 360K).

You can do a similar math for the other formats. Of course, the greater number of tracks and sectors on the higher formats requires that more bytes be used as markers. A 1.44M floppy has 80 tracks × 18 sectors × 512 bytes × 2 sides = 1,474,560 bytes. During formatting, 16896 bytes are used to lay out and mark the tracks and sectors. You end up with 1,457,664 usable bytes (or, rounded off, 1.44M).

Allocation units

The older terminology was cluster, but allocation unit is a better term. On the 360K and 720K disks, DOS arranges for two sectors to be a single allocation unit or cluster. On the 1.2M and 1.44M, however, a single 512-byte sector is an allocation unit. (On hard disk systems, an allocation unit may be made up of four sectors and up to as many as 16).

Parts of two different files cannot be written (i.e., placed in) a single allocation unit because the bytes from the two files would become mixed. The FAT would become very confused if it had to try to separate data from two different files in the same allocation unit. On a 360K or 720K disk, an allocation unit is two sectors or 1024 bytes. If a file is only two bytes, it will require a whole allocation unit.

FAT

The location of each track and sector number is stored on track 0 in a file allocation table (FAT). Whenever a file is written on the disk, it is broken up into 512-

byte chunks. Each of those chunks is written to the first empty sector or allocation unit found.

If the file is long, several allocation units will be required. The system will search for empty units and fill them. Thus, parts of the file might be scattered all over the disk. For instance, some might be on track 5, sectors 8 and 9, and on track 20, sectors 5 and 6. It doesn't matter where the data is because the FAT keeps track of where each part is located. It can then electronically send the head to those tracks and sectors to read all of the file. If you add or erase any part of the file, the FAT will be updated accordingly. Because the tracks and sectors are numbered, the FAT can cause the heads to find any data anywhere on the disk.

The File Allocation Table might be compared to the table of contents in a book.

Directory limitations

DOS sets aside a limited amount of space for the number of files that can be held in the root directory of a floppy disk. (Refer to Table 5-1). For a 360K disk, this amount is 112; for any of the 80-track disks, this maximum number is 224.

Table 5-1 Capacities of various drive types.

Drive type	Tracks/side	Sectors/track	Unformatted capacity	System use	Available to user	Max. dirs.
360K	40	9	368640	6144	362496K	112
1.2M	80	15	1228800	14898	1213952K	224
3.5	80	9	737280	12800	724480K	224
3.5	80	18	1474560	16896	1457664K	224
3.5	80	36	2949120		2.88M	224

You can put many more files on a disk than the listed number. For example, I always keep a copy of all my letters, most of which are only a page long. I have over 500 of them on my hard disk. When I tried to copy them to a 1.44M 3.5" disk, however, I got a file creation error message after number 224 was recorded. Strangely enough, a disk check of the floppy revealed that there was over 600K of empty space on the disk. To avoid the problem, I erased all of the files and then created two separate directories on the disk. Then I was easily able to copy all 500 of my letters into these two directories.

This wasn't difficult: multiple directories can be created on a disk in the same manner as on a hard disk. At the B prompt, I simply typed MD OLDLTRS. (The MD is the command for Make Directory). I then made a second directory called NEWLTRS. I used XTREE to sort my letters according to the date they were written, tagged them, and then copied them to the appropriate directory on the floppy.

Cylinders

The tracks on each side of the disk are directly opposite of each other; in essence, track 0 on side 0 (the top side) is exactly opposite track 0 on side 1 (the bottom side). If you could strip away all of the other tracks, the 0 track on the top and the 0 track on the bottom side might look somewhat like a cylinder, even though it would be rather flat.

Heads

You have certainly heard the old saying that "Two heads are better than one". That is certainly true when it comes to floppy drives. The original drives had a single head. Disks were sold as Single Sided/Double Density (SS/DD). Many people discovered that you could cut a notch on the disk cover, turn the disk over, and then record on the other side. (Many people are doing something similar with the 3.5" 720K floppies; by cutting a hole in the right rear corner, the disks can be used as 1.44M floppies.)

Floppy drives with two heads were soon available. In 1984, an IBM drive with two heads sold for $425. Many people could not afford this high price, so the lower cost single-sided drives were used for some time until the double-sided drives came down in price.

Two heads read and write on the disk: head 0 on the top, and head 1 on the bottom. When head 0 is over track 1, sector 1, on the top of the disk, head 1 is addressing track 1, sector 1 on the bottom side of the disk. The heads move from track to track as a single unit. Data is written to track 1 on the top side, and then the heads are electronically switched to the bottom side and writing is continued to track 1 on the bottom side. Switching between the heads electronically is much faster than moving them to a different track.

TPI

The 40 tracks of a 360K is laid down at a rate of 48 tracks per inch (TPI), so each of the 40 tracks is $1/48$ of an inch wide. The 80 tracks of the high density 1.2M is laid down at a rate of 96 TPI, so each track is $1/96$ of an inch. The 80 tracks of the 3.5" disks are laid down at a density of 135 per inch ($1/135$ of an inch) per track.

Read accuracy

The 5.25" disks have a 1.125" center hole. The drives have a conical spindle that comes up through the hole when the drive latch is closed. This centers the disk so that the heads will be able to find each track. The plastic material of which the disk is made is subject to environmental changes and wear-and-tear. The conical spindle might not center each disk exactly, so head-to-track accuracy is difficult with more than 80 tracks. Most of the 360K disks use a reinforcement hub ring, but it probably doesn't help much. The 1.2M floppies don't use a hub ring. (Coincidentally, except for the hub ring, the 360K and 1.2M disk look exactly the same.)

The tracks of the 3.5" floppies are narrower and greater in density per inch. Because of the metal hub, however, the head tracking accuracy is much better than that of the 5.25" systems.

Hard disks have very accurate head tracking systems. Some have a density of well over 1000 tracks per inch, so much more data can be stored on a hard disk.

Rotation speed

The floppy disks have a very smooth lubricated surface and rotate at a fairly slow 300 RPMs. Magnetic lines of force deteriorate very fast with distance. The closer the heads, the better they can read and write. Thus, the heads directly contact the floppy disks.

In contrast, hard disks rotate at 3600 RPMs. The heads and surface would be severely damaged if they came in contact at this speed. Thus, the heads merely "fly" at a few millionths of an inch above the surface.

The 5.25" standard

Besides different formats, several sizes of floppies used to be sold. The 5.25" became the standard size, however. The early IBM standard divided a single side of the disk into 40 concentric tracks, with each track being divided into 8 sectors for a total of 160K. Then the double-sided disk was developed, so we had a disk that could store a whopping 320K. Later, the 40 tracks were divided into 9 sectors per track, for a total of 180K per side or 360K per disk.

Still later IBM developed the 1.2M floppy: it had 80 concentric tracks divided into 15 sectors per track, which gave us 1.2M per disk.

What is really fantastic is that the 1.2M drives can also read and write to the 360K format. About 40 million PCs are being used, and over half of them have the old 360K drives. Most of the software companies still distribute software on 360K disks because almost everybody can read that format.

The 360K and 1.2M look exactly alike except for the hub ring on the 360K. Still, there is a large difference in the magnetic materials that determines the oersted (Oe) of each one. Oe is a measure of the resistance of a material to being magnetized. The lower the Oe, the easier it is to be magnetized. The 360K has an Oe of 300, while the 1.2M is 600 Oe. The 360K disks are fairly easy to write to, so they require a fairly low head current. The 1.2M is more difficult to magnetize, so a much higher head current is required. This current is switched to match whatever type of disk you tell the system you are using.

You can format a 360K as a 1.2M, but it will find several bad sectors—especially near the center, where the sectors are shorter. These sectors will be marked and locked out. The system might report that you have over 1M of space on a 360K disk, but I don't recommend that you use such a disk for any important data: the data might eventually deteriorate and become unusable.

IBM created a new 3.5" floppy disk standard with the introduction of the PS/2 systems. The 3.5" drives have several good features. The 720K disk can store twice as much data as a 360K, and the 1.44M can store four times as much in a smaller space. They have a hard plastic protective shell, so they are not easily damaged. They also have a spring-loaded shutter that automatically covers and protects the head opening when the disks aren't in use.

The 3.5" drives are much smaller than the 5.25" drives. Thus, they need an adapter in order to mount them in a 5.25" bay. Figure 5-1 and Fig. 5-2 show the 5.25" and 3.5" drives.

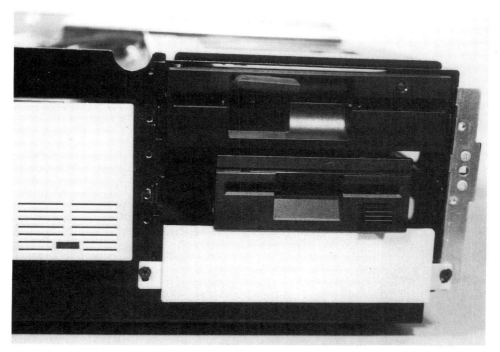
5-1 A 5.25" drive with a 3.5" drive that hasn't had adapters installed.

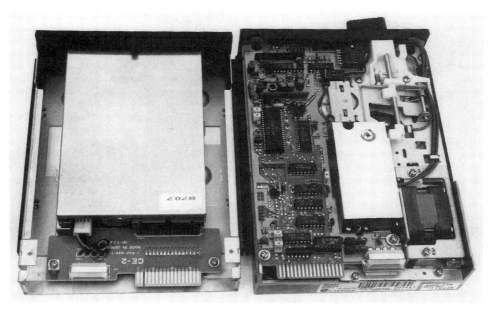
5-2 A 3.5" floppy that has been mounted in the adapter kit alongside a 5.25" drive.

Another feature is the write-protect system. A plastic slide can be moved to open or close a small square hole in the shell. When the slide covers the opening, the disk is write-enabled. When the slide is moved to open the square hole, it is write-protected. This system of write protection is exactly opposite of that used in the 5.25" system.

If the square notch on the 5.25" system is left uncovered, light can shine through, which allows the disk to be written on or erased. If the notch is covered with opaque tape, it can only be read.

Note: Do not use **clear** tape when write-protecting your 5.25" disk. Because the system uses light to verify the protection status, the disk will still seem to be unprotected because the light can shine through the tape. Thus, your disk can still be written on or erased, and you could be in for a nasty surprise at some point.

Figure 5-3 shows four 3.5" disks. The upper two marked MFD2HD (Micro Floppy Disk—the "2" meaning double-sided and the "HD" representing High Density) are high density. The two in the lower part of the photo are double density

5-3 3.5" floppy diskettes. The two on the top are high-density 1.44M diskettes, while the two lower ones are double-density 720K diskettes.

disks, the one on the left being marked as MFD-2DD (Micro Floppy Disk, 2-sided, double-density). Note that the one on the right has no markings at all. Also notice that the arrows at the left top portion of the disks indicate how they should be inserted into the drive. Because of their design, they cannot be completely inserted into a drive upside down or backwards, so no damage can be done that way.

Figure 5-4 shows the reverse side of the disks. Note that the two on the left have the sliding write-protect switch set to enable, while the two on the right are

5-4 The back side of the diskettes. Note the metal hub. The diskette on the top right has the cover pulled open to show the head opening. The diskettes on the right have the holes in the lower-right corner open for write-protect, while the two on the left are closed for write-enable.

open or protected. In order to show the head openings on the two on the right, I have taped the spring loaded shutters so that they remain open.

Formatting

Assuming that the 1.2M is the A drive and you want to format a 360K disk with the 1.2M drive, type

FORMAT A: /4

To format to 1.2M, you need the high density disks. If the system is configured and the controller allows it, you only have to type

FORMAT A:

If you insert a 360K disk, it will try to format it to 1.2M but will probably find several bad sectors.

To format a 720K disk on a 1.44M drive, type

FORMAT B: /T:80 /N:9

To format a 1.44M disk, just type

FORMAT B:

or possibly

FORMAT B: /T:80 /N:18

Helpful .BAT files

I'll show you some batch files that have saved me a lot of time in formatting disks. Here is how I made my batch files:

```
COPY CON FM36.BAT
C: FORMAT A: /4
^Z

COPY CON FM12.BAT
C: FORMAT A: /T:80 /N:15
^Z

COPY CON FM72.BAT
C: FORMAT B: /T:80 /N:9
^Z

COPY CON FM14.BAT
C: FORMAT B: /T:80 /N:18
^Z
```

The ^Z is made by pressing F6. With these batch files, I only have to type fm36 for a 360K, fm12 for a 1.2M, fm72 for a 720K, or fm14 to format a 1.44M.

These statements and .BAT files apply to DR DOS 6.0 and versions of MS-DOS up through v4.01. MS-DOS 5.0 will also format a 360K in a high density drive unless you tell it differently. Still, they have changed the command by adding a switch that can be invoked with the FORMAT command, eliminating the need to type in parameters for the various formats. Just follow these parameters:

Disk Type	Enter this:
360K	FORMAT A:/f:360
12M	FORMAT A:/f:1.2
720K	FORMAT B:/f:720
1.44M	FORMAT B:/f:1.44

All of these commands could be considerably shortened if made into .BAT files similar to those I've shown.

If you try to reformat a previously formatted disk with MS-DOS 5.0, MS-DOS will try to format it the same way. For instance, if you made a mistake and formatted a 360K as a 1.2M, it would insist on formatting it as a 1.2M again if you don't use the /f:360 switch.

Converting a 720K to a 1.44M disk

At one time, the high density 1.44M disks cost as much as $5 per disk. At the same time, the 720K disks were selling for less than $1. You can format and use a 720K as a 1.44M, but disk manufacturers strongly recommend against it.

The Oe of the 720K and the 1.44M is fairly close, and several companies have developed tools for punching the extra hole in 720Ks. These companies claim that the 720K works fine as the higher density disk. Accordingly, they have sold thousands of the punches at about $40 each. Still, if you want to convert a 720K to 1.44M, you don't need a $40 punch. I'll tell you how to convert them yourself for free.

The 720K disks have a plastic cover or shell made from two flat pieces bonded together around the edges. This plastic is rather soft, making it easier for you to penetrate. To mark the location of the hole, take two disks and open the write-protect slide. Place them with the metal hubs facing, and then mark the location through the open write-protect hole. You can use a soldering iron to burn a hole, or else use a pocket knife to dig out a small dimple. The hole or dimple needs to only penetrate through the bottom layer of the plastic cover (which is the side with the metal hub). The hole doesn't have to go all the way through both layers, and it's location or depth isn't too critical. Make the hole and try it. If it doesn't work, enlarge it a bit.

As a test, I converted several 720K disks, formatted them to 1.44M, and then loaded data onto them. Although it's been over a year now, so far I haven't had any trouble. Yet again, the cost of the high density disks is now very reasonable, so I would not recommend taking chances when storing any critical data.

The upper two disks in Fig. 5-5 were 720K that have been converted to 1.44M by drilling a hole in the left rear corner of each one.

5-5 The upper two floppy disks were 720K that have been converted to 1.44M by drilling a hole in the left rear corner of each one. The bottom left is 720K, while the bottom right is 1.44M.

Cost of disks

All floppy disks are now quite reasonable. The 360K DS/DD disks are selling for as low as 21 cents each, and the 720K disks are going for as little as 35 cents each. The 1.2M HD disks are selling at discount houses for as little as 39 cents apiece, while the 1.44M HD disks are selling for as little as 59 cents each.

These prices are real bargains. You can buy ten of the 1.44M disks (i.e., 14.4M of storage) for only $5.90, which means you've only paid about 40 cents per megabyte. Likewise, you can buy ten of the 1.2M HD disks for only $3.90, which means you can store 12M of data on them at a cost of around 32 cents per megabyte. If you use a good compression backup software such as Norton or Fastback Plus, you could store over 28M on the ten 1.44M disks or about 24M on the 1.2M disks.

Discount disk sources

Here are a few companies that sell disks at a discount; of course, several other companies do too. Check computer magazines for ads.

MEI/Micro Center	(800) 634-3478
The Disk Barn	(800) 727-3475
America Group	(800) 288-8025
MidWest Micro	(800) 423-8215

Floppy controllers

A floppy disk must have a controller. In the early days, it was a separate board full of chips. Now, however, they are usually built into a single VLSI chip integrated with a hard disk controller or IDE interface. It might also be integrated with a multifunction board or be built-in on the motherboard.

Higher density systems

The floppy technology continues to advance. Several new higher capacity drives and disks are now available.

Extended density drives

Several companies are now offering a 3.5" Extended Density 2.8M floppy drive. The 2.8M disks have a barium ferrite media and use perpendicular recording to achieve the extended density. In standard recording, the particles are magnetized so that they lay horizontally in the media. In perpendicular recording, however, the particles are stood vertically on end for greater density.

The ED drives are downward-compatible and can read and write to the 720K and 1.44M disks. At the present time, however, they are still rather expensive. IBM has announced that they will use them on one of their PS/2 units, thus probably making these drives the new standard. By the time you read this, they should be fairly reasonable in price.

Very high density drives

Brier Technology—(408) 435-8463—and Insite—(408) 946-8080—have developed 3.5" drives that can store over 20M on a disk. There is no standard among the competing systems, so they use different methods to achieve the very high density.

One of the problems that had to be overcome in very high density drives was that of tracking. The drives have little trouble reading and writing to the 135 tracks per inch (TPI) of the standard 3.5" disk. Still, 20M requires many more much closer tracks.

Brier Technology's Flextra uses special disks that have special magnetic servo tracks embedded beneath the data tracks. The Insite disks have optical servo tracks that have been etched into the surface with a laser beam. The heads then lock onto the servo tracks for accurate reading and writing to the data tracks. The Insite drive has a head with two different gaps, which allows it to read and write to the 20M format as well as the 720K and 1.44M formats.

The Brier drive is being distributed by the Q'COR Company—(800) 548-3420. You can contact Insite at (408) 946-8080. The special disks for these systems cost about $20 each.

Bernoulli drives

The IOMEGA Corporation has a high capacity Bernoulli floppy disk system. Their system allows the recording of up to 44M on a special floppy disk. The Bernoulli disks spin much faster than a standard floppy, which forces the flexible disk to bend around the heads without actually touching them. This is in accordance with the principle discovered by the Swiss scientist, Jakob Bernoulli (1654–1705).

The average seek time for the Bernoulli systems is 32 ms, while the better hard drives have about 15 ms.

Bernoulli drives are ideal for areas where the data may be confidential. Each person in an office may have their own 44M floppy that can be removed and locked up. This system is also great for backing up a hard disk system.

IOMEGA has had the field to themselves for several years, so the Bernoulli box has always been a bit expensive (at about $1500 for a drive and about $70 for each disk). Brier Technology, Insite, and several other companies, however, are now giving them some competition with their very high density systems. We should soon see prices that are quite reasonable on the very high density floppy disk systems.

Mounting a 3.5" drive

Figure 5-1 shows a 3.5" floppy without an adapter sitting in a drive bay opening below a 5.25" drive. Figure 5-2 shows the 3.5" drive with the extenders and adapters alongside the 5.25" floppy. There are special extender frames that fit around the 3.5" drives, allowing them to be mounted in the standard 5.25" openings. Check your case: many of the newer ones have bays for a 3.5", but some might have to be mounted on their side in the bay.

What to buy

Many vendors are still advertising and selling 360K and 720K drives, but I don't know why anyone should buy one. They are obsolete. I would recommend either the 1.2M and 1.44M drives, or possibly the Extended Density 2.88M drives.

If you live near a large city, you should have lots of stores nearby, as well as computer shows and swap meets. If you don't live near a good source, then your next best bet would be a mail order house.

Lots of computer magazines are full of ads. Space for these ads is rather expensive, so quite often the vendors use abbreviations. In case you have trouble understanding the ads, floppy disk drives are usually listed as FDD and floppy drive controllers are FDC. You must read the ads carefully.

Chapter 6
Hard disks and mass storage

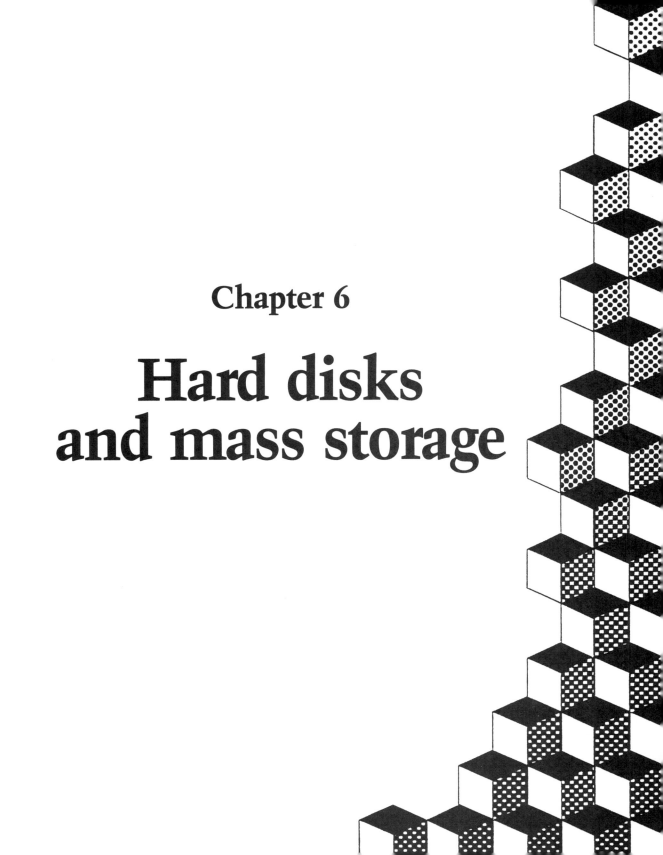

You could buy a computer with just one or two floppy drives and without a hard disk if you wanted. At one time, floppy disks were all that anyone had. In addition, software programs were small and usually fit on one floppy. For example, when I got my first copy of WordStar 3.0, the entire program was on a 140K single-sided disk. Today, though, WordStar 6.0 comes on 21 360K diskettes. Most modern software programs require from 2 – 5M of disk space: there's no way you could run some of them from a floppy disk.

You might be able to save a few dollars by not buying a hard disk; but if your time is worth anything at all and you do any serious computing, you will waste far more time than the cost of a hard disk. Besides, hard disks aren't really that expensive.

One reason to own a 386 is because it can handle lots of memory. In virtual mode, it can address the data that could be stored on 32,000 20M hard disks (i.e., 640,000,000,000 bytes). If you are building a 386 for your own personal use at home, you probably won't ever need that capability. For large businesses, companies, and factories, however, it is quite often necessary to be able to address large amounts of memory such as employee payroll files, customer lists, accounting, databases, spreadsheets, and maybe even the boss's golf scores.

Another reason to own a hard disk is for convenience. Years ago, before I got my first hard disk, I often spent hours looking for a certain floppy disk. I now own several computers, some with 100M hard disks and some with 200M. I can find and load any one of hundreds of files within seconds with just a few keystrokes; they are no further away than the tips of my fingers.

There is an intoxicating feeling of power in knowing that you have several hundred programs at your fingertips that can pop up on your screen within milliseconds. Of course, you still need to keep the floppies as a backup just in case something happens to your hard disk. In many cases, you might not ever need the majority of the software that you accumulate. As always, though, there is some kind of immutable and inflexible law that decrees that if you throw something away, you will almost certainly need it the next day.

Another natural law decrees that the requirements for disk storage will expand to fill whatever space is available. This theory is especially valid if the computer is in an office where several people use it or if it is used as a multiuser or server on a Local Area Network (LAN). I am the only one who uses my computers with 100M disks, but these disks are filling up so fast it almost seems as if those little bytes are getting together reproducing themselves.

Choosing a hard disk

You must consider several factors before deciding to buy a hard disk. Of course, the primary factors will be what you need to do with your computer and how much you want to spend on it.

The next sections briefly describe some of the factors that should influence your buying decision.

Capacity

Buy the biggest drive you can afford; no matter how big it is, it will soon be filled up. Don't even think of buying anything less than 40M; better yet, get at least 80M. New software programs have become more and more friendly and offer more and more options—and thus require more and more disk space.

Speed or access time

Access time is the time it takes a hard disk to locate and retrieve a sector of data. This amount includes the time that it takes to move the head to the track, settle down, and read the data. For a high-end, very fast disk, this might be as little as 9 milliseconds (ms). Some of the older drives and systems required as much as 100 ms. An 85 ms hard drive might be okay for a slow XT, but I would recommend the fastest you can afford. Depending on what you are using it for, a 28 ms drive might not be fast enough for a 386SX; a 16 ms would be better.

Of course, the faster the hard disk, the more expensive it will be. Thus, your choice here will depend a lot on your computing needs and your spending habits.

Type of drive, stepper, or voice coil

Most of the less expensive hard drives use a *stepper motor*, which moves the heads in discrete steps (increments) across the disk until they are over the track to be read or written. You can usually hear the heads as they move from track to track.

The voice coil hard drives are quieter, a bit faster, and more reliable—and, of course, more expensive. There might not be any marking on the drive to indicate whether it is a voice coil or not, but their spec sheets will show that they have an odd number of heads. Actually they have an even number of heads—one on the top and bottom of each platter. However, one head and platter surface is used as a servo control for the other heads. The servo head follows a tracking system recorded by the factory. When a program calls for data to be read from a specific track, the servo head moves to that track very quickly and accurately. Because all of the heads move as one, the head designated to read the data is also moved to the proper track.

A voice coil system is very quiet; unless there is an LED indicator on the front panel, you might not even realize that it's operating.

A bit of history

Incidentally, you might have seen the term *fixed disks* in IBM DOS manuals and other IBM literature. Almost everybody else in the computer world calls them hard disks. Still, because IBM was one of the pioneer developers of the hard disk, I suppose they can call them any name they want.

One of the first hard disk drive systems was developed by IBM. The system had a large 30M hard disk that could be removed, as well as a 30M fixed internal hard disk.

The Winchester House, a tourist attraction, is located in San Jose on Winchester Boulevard not far from an IBM plant. This house was built by the widow of the famous inventor of the Winchester .30/.30 rifle. Because the IBM hard disk was a 30/30 system, someone hung the name Winchester on it; for reasons unknown, it stuck and now all hard disks that use that original technology are known as Winchesters.

IBM also refers to the Winchester as a Direct Access Storage Device, abbreviated DASD (and pronounced *DAZ-dee*).

What size disk do you need?

Just a few years ago, a 10M hard disk was sufficient. Now 80M might not be enough. The more popular user friendly programs have lots of on-screen help and many menus. Some of the programs such as Paradox, Windows, WordStar, WordPerfect, Excel and others can use up the better part of 80M in a hurry. If you load in a few of these programs, you won't have much workspace left or room to install other goodies.

We, the end users, are fortunate in that there is a lot of competition in the manufacturing of hard disks. Like most of the other computer components, the prices of hard disks continue to come down. Just a few years ago, a 20M hard disk system cost over $2000. You can now buy a good 80M hard disk for about $250. A standard controller might cost less than $50.

How a hard disk operates

Basically the hard disk resembles the floppy. It spins a disk coated (plated) with a magnetic media. The hard disks are also formatted similarly to the floppy. However, the 360K floppy disks have only 40 tracks per inch (TPI), while the hard disks might have anywhere from 300 to 2400 TPI. The 360K floppy has 9 sectors per track, while the hard disk might have from 17 to 54 sectors per track. Both floppy and hard disks store 512 bytes per sector, though.

Another major difference is the speed of rotation. A floppy disk rotates at about 300 RPM, while a hard disk rotates at 3600 RPM.

Leaving it turned on

Incidentally, another difference in the hard disk and the floppy is that the floppy comes on only when it is needed. Because of its mass the hard disk takes quite a while to get up to speed and to stabilize. Thus, it comes on whenever the computer is turned on and spins as long as the computer is on.

If you are going to use your computer several times during the day, many experts have suggested that you leave it on all day. Every time an electronic circuit is turned on, a stream of current rushes through it and strains it. Accordingly, some busy people never turn their computer off.

MTBF

The Mean Time Before Failure (MTBF) is a figure representing how long the average drive should last before it fails and is determined by the manufacturer. This number should give you an idea of the reliability and quality of the drive, with the higher MTBF representing the more durable drive. Most manufacturers claim a MTBF of 40,000 to 50,000 hours.

I have seen some drives advertised with a MTBF of 150,000 hours. If the computer was turned on 8 hours a day, 7 days a week, this means that it should last for 18,750 days, or over 51 years. I don't know how the manufacturer could have determined this figure, nor can I quite believe it. I even find it difficult to trust the 50,000 hour figure, which would take 17 years at 8 hours a day.

The MTBF is just an average, somewhat like the average life of a man (estimated at 73 years at this time). However, some males die in infancy, while a few others live to be 100. Similarly, hard drive life will vary.

Reading and writing

Everything that a computer does depends on precise timing. Crystals and oscillators are set up so that certain circuits perform a task at a specific time. These oscillating circuits are usually called *clock circuits*.

You have probably seen representations of magnetic lines of force around a magnet. To write data to the hard disk, electrical on and off pulses are sent to the head. These pulses magnetize small sections of the track to represent 0s and 1s. The magnetized spot on a disk track has similar lines of force.

To read the data on the disk, the head is positioned over the track and the lines of force from each magnetized area cause a pulse of voltage to be induced in the head. During a precise block of time, an induced pulse of voltage can represent a 1 and the lack of a pulse can represent a 0. The amount of magnetism induced on a disk when it is recorded is very small. It must be small so that it will not affect other tracks on each side of it or affect the tracks on the other side of the disk. Magnetic lines of force decrease as you move away from a magnet by the square of the distance, so the heads must be as close to the disk as possible.

The floppy disk heads actually contact the diskette, which causes some wear but not much because of the fairly slow rotation. The plastic floppy disks also have a special lubricant and are fairly slippery. However, heads of the hard disk systems never touch the disk; the fragile heads and the disk would be severely damaged if they would make contact at the scorching speed of 3600 RPMs. Instead, the heads hover over the spinning disk, just micro-inches above it. Accordingly, the air must be filtered and pure because the smallest speck of dust or dirt can cause the head to crash.

Platters

The surface of the hard disk platters must be very smooth. Because the heads are only a few millionths of an inch away from the surface, any unevenness could cause a head crash. The hard disk platters are usually made from aluminum (which is nonmagnetic) and lapped to a mirror finish. Some companies are now using a glass substrate. Once they have the smooth mirror finish, the disks are coated or plated with a magnetic material.

The platters also must be very rigid so that the close distance between the head and the platter surface is maintained. You should avoid any sudden movement of the computer or any jarring while the disk is spinning because it could cause the head to crash onto the disk and damage it. Most of the newer hard disk systems automatically move the heads to the center, away from the read/write surface, when the power is turned off.

A hard disk system might have only 1 platter or as many as 10 or more. All of the platters (disks) are stacked on a single shaft with just enough spacing between each one for the heads.

Head positioners

Several different types of head positioners are available. Some use stepper motors that move the heads in discrete steps and position them over a specified track. Others use a worm gear or screw-type shaft that moves the heads in and out.

The faster drive systems use voice coil technology similar to that used for loudspeakers in hi-fi systems. The voice coil of a loudspeaker is made up of a coil of wire wound on a hollow tube attached to the material of the speaker cone. Permanent magnets are then placed inside or outside the coil. Whenever a voltage is passed through the coil of wire, it will cause magnetic lines of force to be built up around the coil. Depending on the polarity of the input voltage, these lines of magnetic flux will be either the same or opposite of the lines of force of the permanent magnets.

If the polarity of the voltage, for instance a plus voltage, causes the lines of force to be the same as the permanent magnet, then they will repel each other and the voice coil might move forward. If they are opposite, they will attract each other and the coil will move backwards. The voice coil system can move the heads quickly and smoothly to the desired track area.

Drive systems
MFM

Most of the early hard disk systems used the *Modified Frequency Modulation (MFM)* method. This system formats each track into 17 sectors; and, just like the floppy disk system, 512 bytes can be stored in each sector. So if we multiply 512 bytes × 17 sectors, we can see that 8704 bytes can be stored on each track.

The clock frequency for the standard Modified Frequency Modulation (MFM) method of reading and writing to a hard disk is 10MHz per second. To write on the disk during one second, the voltage might go high for the first $1/100$ of a second, then turn off for the next $1/100$ of a second, then back on for a certain length of time. The track on the spinning disk is moving at a constant speed beneath the head. Blocks of data are written or read during the precise timing of the clock.

Because the voltage must go plus or zero, in order to write 1s and 0s, the maximum data transfer rate is only 5 megabits per second, just half of the clock frequency.

Most of the MFM systems are reliable but slow, limited in capacity, and bulky in size. Now they are practically obsolete. If you have more time than money, however, they are relatively inexpensive.

RLL

The *Run Length Limited (RLL)* system was developed by IBM some time ago for use on large mainframe hard disks. Adaptec Corporation of Milpitas first adapted the technology so that it could be used on PC hard disk controllers. This system divides each track into 26 sectors with 512 bytes in each sector, or 26 × 512 = 13,312 bytes per track. This is over 50 percent more than the 8704 per track allowed by the MFM system.

The RLL system also has a transfer rate 50% higher than MFM at 7.5 megabits per second.

IDE

The *Integrated Drive Electronics (IDE)* have the controller electronics built onto the drive. This allows the manufacturer to optimize the drive's capabilities. They do need an interface to the computer bus. Many motherboards now have this interface built-in. If it does not, you will need an interface to plug into one of your slots. This interface with a floppy controller might cost as little as $20. The IDE interface is also built into some multifunction boards. The IDE drives are usually small, fast, and have a fairly large capacity. Figure 6-1 shows an IDE drive with a single 40-wire ribbon cable.

6-1 An IDE drive with a single 40-wire ribbon cable.

SCSI

The *Small Computer Systems Interface (SCSI)*, pronounced "scuzzy," also has its controller built onto the drive. They need an interface, but up to seven other SCSI devices can be connected in a daisy chain to the one interface. The SCSI drives are usually fast and have a high capacity. They can format from 26 to 54 sectors per track. The transfer time can be 10 – 15 megabits per second. Most SCSI drives are used on large high-end systems. Figure 6-2 shows the rear of a SCSI drive with a 50-wire ribbon cable.

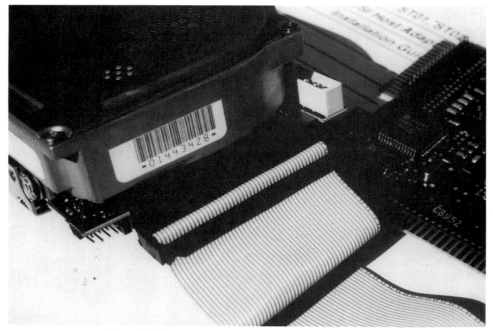

6-2 A SCSI drive with a single 50-wire ribbon cable.

ESDI

The *Enhanced Small Device Interface (ESDI)*, pronounced "ezdy," is a drive system that operates at 10 megabits or more per second. It can format from 32 up to 54 sectors per track and are usually very fast, have a high capacity, and a high cost. They are usually found on high-end systems.

Controllers

A hard disk needs a controller to handle the data going back and forth to the disk. In the early days, electronic boards were expensive and not very sophisticated. Most were so slow that the interleave would often have to be as much as 6 to 1. Thus, it would read one sector and then let six more sectors pass beneath the head while it digested the input data. It would then be ready for the next sector read and then wait for six more sectors to pass by.

To be perfectly fair, this situation wasn't entirely the controller's fault. The early computers and disk drives were very slow compared to what we have today.

The controller boards were usually manufactured by someone other than the disk drive manufacturer. It was sometimes very difficult to set up all the switches and jumpers so that the controller would work with a particular drive and the computer system.

Controllers today are inexpensive, fast, and sophisticated. Most of them today will handle 1 to 1 interleave (direct transfer) with no difficulty. Figure 6-3 shows a 1:1 MFM hard disk and floppy disk controller (HDC/FDC).

Often the early controllers controlled the hard disk only. You had to buy a separate controller board for the floppy drives. Now almost all of the hard disk controllers have the floppy controller integrated on the board.

You must buy a controller to match the type of hard disk you own. An MFM drive will require an MFM controller. The same goes for RLL, IDE, SCSI, or ESDI. The controllers can vary greatly in price, from about $20 for an IDE up to several hundred for a high-end SCSI.

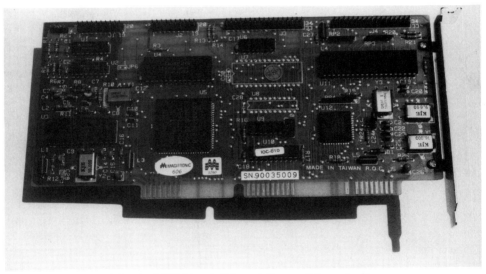

6-3   An MFM hard disk and floppy disk controller.

Formatting a hard disk

Hard disks require two levels of formatting: a low-level and a high-level. In most cases, the low-level formatting is now done at the factory, especially for IDE, ESDI, and SCSI drives. You should not try to low-level format these drives. The documentation that came with your drive should tell you whether it needs to be low-level formatted and how to do it.

Although the disk might have been low-level formatted at the factory, you will probably have to do the high-level formatting. If it is a large disk, you might want to partition it into two or more logical drives. Early versions of DOS limited you to 32M per disk, but it will now let you have several hundred megabytes as a single disk.

Use the DOS FDISK command to partition the disk and then high-level format each partition. DOS allows you to have up to 23 logical hard drives, which can be all of the letters of the alphabet except A, B, and C. Letters A and B are reserved for floppy drives, and C is reserved for the first active hard drive. I would recommend that you partition large disks into two or more logical drives: if you have a crash or disk failure on one logical drive, the data in the other sections might still be recoverable, whereas you might not be able to recover any data from a single crashed logical drive. Of course, you should always have your hard disks backed up.

CMOS ROM setup

The system ROM must be set up to recognize the type of hard disk that you have installed. The system must know the number of heads, cylinders, the capacity, the landing zone for the heads, and several other things. Different model drives and drives from different manufacturers might all be different in some respect. The original IBM system grouped and recognized 15 different types at that time in 1984. Most of today's systems recognize 46 different types. They also list a 47th type where they let you type in the various characteristics if your drive does not fit any of the listed types.

Adding one or more hard drives

Most hard drive controllers and interfaces have provisions for adding a second hard drive. Some systems might require that the second hard drive be the same type, size, and from the same manufacturer, while other systems might not care as long as the second drive was the same general type (e.g., if your controller and first drive is a MFM, then the second drive must be MFM). Most of the drives have some sort of jumper or shorting bar that must be installed to operate the second hard drive. The pen in Fig. 6-4 points to the pins that must be shorted to configure these drives as number 1 and 2.

You can add two SCSI drives to a system that has MFM, RLL, or IDE drives. Thus, you could have two SCSI drives and two of the other types. You cannot mix MFM, RLL, or IDE drives, however; this would cause a conflict of the I/O interrupt system. Most of the SCSI drives use a special driver to avoid conflicts.

Hard cards

Several companies make plug-in hard cards. These cards are small, fast, and can have 40–200M. They are very easy to install: just plug them into an empty slot on the motherboard. The cards have no cables or controllers to worry about and usu-

6-4 The pen points to the black shorting bar that determines which of the two hard disks is number one. Note that the second set of pins are shorted on the bottom disk, making it disk number 2.

ally come with a software driver that avoids I/O conflicts. Still, they are usually a bit more expensive than an equivalent standard type of hard disk.

External drives

Several companies have developed drives that can be plugged into the serial port or a parallel port. They can be used for backup or for downloading and transferring data from one computer to another.

The Pacific Rim Company, (415) 782-1013, has a small battery-powered 20M hard disk that plugs into the parallel port. I bought a Toshiba laptop without a hard disk and soon realized that I had made a very big mistake. Even with 1.44M floppies, I found it difficult to do any productive work on it. However, it works great with the Pacific Rim hard drive (see Fig. 6-5).

Compression

One way to add more space to your hard disk is to use compression, which has been around for several years. Most bulletin boards use one form of compression

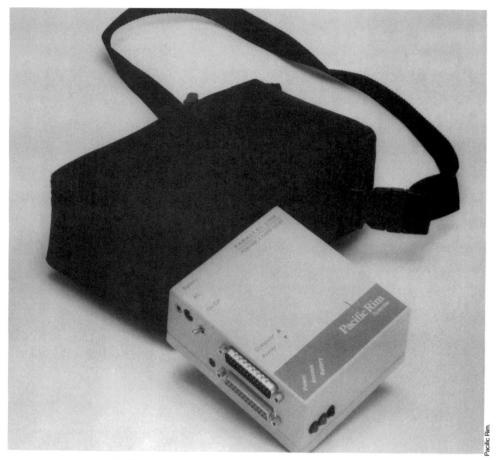

6-5 The Pacific Rim battery-powered 20M hard disk.

or another so that they can save cost in storing the data and in telephone and modem costs. You could compress a file by half or more.

Most of the larger software companies are now shipping their products in a compressed form, many of them under PKZip format. They are unzipped (i.e., expanded) when you load them on your hard disk.

In the past, one problem with compression was that programs took a lot of time to be compressed and then expanded. However, several companies today offer both software and hardware compression systems almost as fast as the fastest hard disk. One that I've been using for some time is Stacker, from Stac Electronics, (619) 431-7474. It can almost double the amount of space on any hard disk.

Stacker has both a software and plug-in board, or a software-only version. The software-alone version lists for $99 and has the capability to double the room in a 100M hard disk. You cannot buy a hard disk for that amount. Using the board version (listing at $199) makes it just a bit faster.

I set up the software version on a 30M disk, so now when I do a CHKDSK, I'm told that I have 60M of usable space. When I install it on the 20M Pacific Rim for my laptop, I get 40M of space. If I didn't know that Stacker was there, I wouldn't be able to tell the difference. I've found it to be fast and reliable.

Figure 6-6 shows the Stacker board and software.

6-6 The Stacker Compression board and software.

CD-ROM

The Compact Disc Read-Only Memory industry is one of the fastest growing of all the computer peripherals. Sony, Hitachi, Phillips, JVC, Amdek, Panasonic, and several others are manufacturing the drives, which are all compatible and can be interfaced to a 386 with a plug-in board.

Since that time when the first stone age man picked up a charred stick from his fire and drew a picture on the cave wall, man has been searching for newer and better ways to record data. There have been several noteworthy achievements—chiseled stone, writing on clay, then on papyrus, then Gutenberg with his press and the first printed Bible.

Today we have mountains of printed matter as well as a vast amount of data on magnetic disks and optical systems. We are almost inundated in a sea of information. Each day, the haystacks become magnitudes larger and the needle becomes more and more difficult to find. The emerging CD-ROM technology simplifies sifting through the mountains of information to find desired information.

These mountains of information are going to become even larger. Thousands of CD-ROM disks are available today. The off-the-shelf disks available cover a wide variety of subjects such as library and bookstore reference materials, general reference, literature, and art. Disks contain information on business, biology, medicine, physics, and most other branches of science and technology, as well as law, finances, geology, geography, government, and many other topics. In time, disks on almost any subject in a library will be marketed.

Several companies manufacture half-height CD-ROM drives similar to the one shown in Fig. 6-7. This one was built by the Hitachi Corporation—(800) 262-1502.

6-7 A half-height CD-ROM drive.

The High Sierra Standard

One reason for the growth of the CD-ROM industry is that most of the larger companies have cooperated to form a standard. A group met at the High Sierra Hotel in Lake Tahoe (explaining the standard's name of High Sierra). Of course some companies have not adopted the standard, but almost all of the off-the-shelf disks will run on any of the several drives available.

Microsoft, creator of MS-DOS, was one of the participants in creating the High Sierra Standard. They developed an MS-DOS Extension software package that is a part of the standard. You can store over 600M on a CD-ROM disk (while a floppy diskette holds only 360K bytes). If we divide 600,000,000 by 360,000, we find that we could store the contents of over 1,666 floppy disks on a single side of a CD-ROM.

One of the first CD-ROM disks published had the 21 volumes of Grolier's Encyclopedia on it. These 21 volumes take up less than 20% of a single side of the 12 cm disk (4.72"). Any significant word, phrase, or subject can be found and displayed on the monitor almost instantaneously. Searches that might take hours in the printed volumes can be done in just seconds. Microsoft Bookshelf is a popular and useful disk, including Roget's Thesaurus, American Heritage Dictionary, and eight other books essential to anyone who does any writing.

Another very popular disk is one published by PC-SIG, (800) 245-6717, and containing over 700 public domain programs.

Chapter 7

Backup

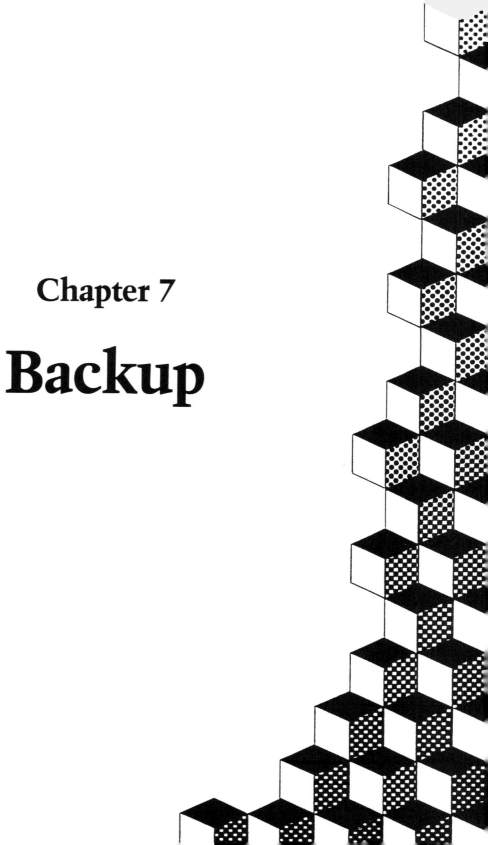

If you buy a software program, the very first thing you should do is write-protect the floppies. You can very easily become distracted and write on a program disk in error, which would probably ruin the program. The vendor might give you a new copy, but you'd still probably have to wait for weeks and fill out paperwork.

If you are using 5.25" floppies, you should cover the square write-protect notch with a piece of opaque tape (Scotch or clear tape won't work). The drive focuses a light through the square notch. If the light detector can sense the light, it will allow the disk to be written on, read, or erased. If the notch is covered with opaque tape, the disk can be read but not written on or erased. Some vendors now distribute their programs on disks without the square notch.

If you are using 3.5" floppies, you should move the small slide on the left rear side so that the square hole is open. The 3.5" write-protect system is just the opposite of the 5.25" system. The 3.5" drive uses a small microswitch to check the square hole. If it is open, the switch will allow the disk to be read but not written on or erased. If the slide is moved to cover the square hole, the disk can be written on, read, or erased.

It takes less than a minute (actually, only a few seconds) to write-protect a disk. These few seconds might save you lots of valuable time later. If a program disk is ruined because it wasn't protected, it might take weeks to get a replacement for the original and you might even have to buy a new copy of the program.

After you have made sure that the disks are write-protected, you should use DISKCOPY to make exact copies of your original disks. The originals should then be stored away, with you using only the copies. Then if you damage one of them, you can always make another copy from the original.

Unerase software

Anyone who works with computers for any length of time is bound to make a few errors. One of the best protections against errors is to have a backup. The second best protection is to have a good utility program such as Norton Utilities or PC Tools. These programs can unerase a file or even unformat a disk. When a file is erased, DOS goes to the FAT table and deletes the first letter of each filename. All of the data remains on the disk unless a new file is written over it. If you have erased a file in error or formatted a disk in error, *do not do anything to it until you have tried using a recover utility*. Don't use the DOS Recover utility except as a last resort. Use Norton Utilities—(213) 319-2000—or if you are using Windows, use

Norton Desktop for Windows	(408) 253-9600
Mace Utilities	(504) 291-7221
PC Tools	(503) 690-8090
DOSUTILS	(612) 937-1107

or any of several other recovery utilities. All these utilities allow you to restore the files by replacing the missing first letter of the filename.

The early versions of DOS made it very easy to format your hard disk in error. If you happened to be on your hard disk and typed FORMAT, it would immediately begin to format your hard disk and wipe out everything. Later versions will not format unless you specify a drive letter. These versions also allow you to include a volume label (name) on the drive when you format it by including the /v. You also can add a label name later by using the command LABEL. If the drive has a volume label, it cannot be formatted unless the drive letter and correct volume name is specified. (You can display, delete, assign, or change the name of a volume by typing the command LABEL. The label name is also displayed when CHKDSK is run.)

All people, including myself, have erased files in error in the past. Everyone is only human, so we will all do it again at some point. Unfortunately, some of us will not have backups and unerase software. In a fraction of a second, we will wipe out data that might be worth thousands of dollars, hundreds of hours, work impossible to duplicate. Yet many people have not backed up their precious data, most of them having been fortunate enough not to have had a major catastrophe.

Just as sure as we will have more earthquakes in California, you can look forward to at least one unfortunate disaster if you use a computer long enough. Mistakes can be made in thousands of ways; you can't prevent them. If your data is backed up, however, it doesn't have to be a disaster. I'd rather be backed up than sorry.

Jumbled FAT

I talked about the all-important File Allocation Table (FAT) in Chapter 6. This Table keeps a record of the location of all the files on the disk. Parts of a file might be located in several sectors, but the FAT knows exactly where they are. If for some reason track 0 (where the FAT is located) is damaged, erased or becomes defective, you won't be able to read or write to any of the files on the disk.

Because the FAT is so important, programs such as PC TOOLS and Mace Utilities can make a copy of the FAT and store it in another location on the disk. Every time you add a file or edit one, the FAT changes, so these programs make a new copy every time the FAT is altered. If the original FAT is damaged, you can still get your data by using the alternate FAT.

Head crash

The heads of a hard disk "fly" over the disk just a few micro-inches from the surface. They must be close in order to detect the small magnetic changes in the tracks. The disk spins at a dangerous 3600 RPMs; if the heads contact the surface of the fast-spinning disk, they can scratch it and ruin the disk.

A sudden jar or bump to the computer while the hard disk is spinning can cause the heads to crash, as well as can a mechanical failure or some other factor. You should never move or bump your computer while the hard disk is running.

Most of the newer disks have a built-in safe park utility. When the power is removed, the head is automatically moved to the center of the disk where there are no tracks. Some of the older disks do not have this utility, however. You can also crash the head if the power is suddenly removed, such as in a power failure.

The technology of the hard disk systems has improved tremendously over the last couple of years, but they are still mechanical devices. As such, you can be sure that eventually they will wear out, fail, or crash.

Most hard disks are now relatively bug-free. Manufacturers quote figures of several thousand hours mean-time before failure (MTBF), but these figures are only an average. You have no guarantee that a disk won't fail in the next few minutes. A hard disk is made up of several mechanical parts that will wear out and fail if used long enough. Many vendors list MTBF figures of 40,000 to 150,000 hours, which means that disk should last for several years. Still, several businesses do nothing but repair hard disks that have crashed or failed.

Crash recovery

A failure can be frustrating, time-consuming, and make you feel utterly helpless. In the unhappy event of a crash, depending on the severity, you might be able to retrieve some of your data one way or another.

Some companies specialize in recovering data and rebuilding hard disks. Many of them have sophisticated tools and software that can recover some data if the disk is not completely ruined. If it is possible to recover any of the data, the Ontrack Computer Systems at (612) 937-1107 can probably do it. You can find several others in computer magazine ads. A couple of companies that I have used to recover data are the California Disk Drive Repair at (408) 727-2475 and Rotating Memory Service (RMS) at (916) 939-7500. RMS supplied the crashed disk for the photo in Fig. 7-1. They were unable to recover any data from this disk because the damage was much more severe than that in most crashes.

The cost for recovery services can be rather expensive, but if you have critical data, then it's usually worth it. You'll find a simple backup to be much cheaper, though.

Small logical drives are better

Early versions of DOS would not recognize a hard disk larger than 32M, but DOS now allows you to have a C or D drive as large as 512M or more. I warn you, though: *Don't do it*. If this large hard disk crashed, you might not be able to recover any of its data. If the same disk was divided into several smaller logical drives and only one of the logical sections failed, you still might be able to recover data in the unaffected logical drives.

Excuses

Next follows some excuses not to backup and why they're wrong.

7-1 A crashed disk.

"I don't have the time"

This is not a good excuse. If your data is worth anything at all, it's worth backing up. You only need a few minutes to back up a large hard disk with some of the newer software.

"It's too much trouble"

Making backups is a bit of trouble unless you have an expensive tape automated backup system. I admit that backups can require a bit of disk swapping, labeling, and storing. Still, with a little organizing, you can do it easily. If you keep all of the disks together, you don't have to label each one. Just stack them in order, put a rubber band around them, and use one label for the first one of the lot.

Yes, it is a bit of trouble to make backups; but if you don't have a backup, consider the trouble you'd go through to redo the files from a disk that has crashed. The comparative trouble that it takes to make a backup is negligible.

"I don't have the necessary disks, software, or tools"

Depending on the amount of data to be backed up and the software used, you might need 50–100 360K disks. (Of course, using high density disks would decrease the necessary amount of disks.) Again, backup takes only the few minutes and disks to copy only the data that has been changed or altered. In most cases, the same disks can be reused the next day to update the files. Several discount mail order houses sell 360K disks for as little as 25 cents apiece, 39 cents each for 1.2M and 720K, and 59 cents for 1.44Mb. I've listed several discount companies at the end of Chapter 5.

"Failures and disasters only happen to other people"

The only people who believe this are those who have never experienced a disaster. Unfortunately, you won't be able to convince them; they'll just have to learn the hard way.

Additional reasons to backup
General failure

Outside of ordinary care, one can't do much to prevent a general failure. A failure could occur because of a component on the hard disk electronics, in the controller system, or elsewhere. Even things such as a power failure during a read/write operation can cause data corruption.

Theft and burglary

Computers are easy to sell, so they're favorite targets for burglars. It would be bad enough to lose a computer, but many computers have hard disks full of data even more valuable than the computer itself.

Because of theft, you should put your name and address on several of the files on your hard disk. You should also scratch some identifying marks on the back and bottom of the case, as well as write down the serial numbers of your monitor and drives.

Also, store your backup files in an area away from your computer. Thus, there'll be less chance of losing both computer and backups in a burglary or fire.

Archival

You can also backup for archival purposes. No matter how large the hard disk might be, it will eventually fill up with data. Quite often, some files will no longer be used or perhaps only used once in a great while. I keep copies of all the letters that I write (in the hundreds) on disk. Rather than erase the old files or old letters, I just put them on a disk and store them away.

Fragmentation

After a hard disk has been used for some time, files begin to be fragmented—the data is recorded on concentric tracks in separate sectors. If part of a file is erased or changed, some of the data might be in a sector on track 20 and another part on track 40. Also, some sectors on several tracks might be open because portions of data have been erased. Hunting all over the disk for pieces of a file can slow the disk down. If the disk is backed up completely and then erased, the files can be restored so that they will again be recorded in contiguous sectors. The utility programs mentioned above can unfragment a hard disk by copying portions of the disk to memory and rearranging the data in contiguous files.

Data transfer

Often you must transfer a large amount of data from one hard disk on a computer to another, a task made easy and fast by a good backup program. You can easily make several copies that can be distributed to others in the company. This method could be used to distribute data, company policies and procedures, sales figures, and other information to several people in a large office or company. The data could also be easily shipped or mailed to branch offices, customers, or others almost anywhere.

Methods of backup
Software

You can make basically two types of backup—the image backup and the file-oriented backup. An *image backup* is an exact bit-for-bit copy of the hard disk copied as a continuous stream of data. This type of backup is rather inflexible and does not allow for a separate file backup or restoration. Conversely, the *file-oriented* type of backup identifies and indexes each file separately, thus allowing a separate file or directory to be backed up and restored easily. Backing up an entire 40M or more each day can consume a great deal of time; with a file-oriented type system, however, once a full backup has been made, you only need to make incremental backups of those files that have been changed or altered.

DOS stores an archive attribute in each file directory entry. When a file is created, DOS turns the archive attribute flag on. If the file is backed up by using DOS BACKUP or any of the commercial backup programs, the archive attribute flag is turned off. If this file is later altered or changed, DOS will turn the attribute back on. At the next backup, you can have the program search the files and look for the attribute flag. You can then backup only those that have been altered or changed since the last backup. You can view or modify a file's archive attribute by using the DOS ATTRIB command.

Several very good software programs on the market let you use a 5.25" or 3.5" disk drive to backup your data. Again, you should have backups of all your master software so you don't have to worry about backing up that software every day. Because DOS stamps each file with the date and time of creation, you can easily backup only those files that were created after a certain date and time.

Once the first backup is made, all subsequent backups only need to store changed or updated data. Most backup programs can recognize whether a file has been changed since the last backup. Most of them can also look at the date stamped on each file and back up only those within a specified date range. Thus, you should only need a few minutes to copy just those new or changed files. In addition, you normally shouldn't have to backup your program software (because you have the original software disks safely tucked away—don't you?).

BACKUP.COM

One of the least expensive methods of backup is to use the BACKUP.COM and RESTORE.COM that comes with MS-DOS. However, you have a price to pay in that these files are slow, time-consuming, and rather difficult to use. Still, they will do the job if nothing else is available.

Here are just a few commercial backup programs that should work better. Of course, many others are also on the market.

Norton Backup Norton Backup, (213) 319-2000, is one of the newest and fastest backup programs on the market, as well as one of the easiest to use. It compresses the data so that fewer disks are needed.

Norton Desktop for Windows Norton and Symantec have now merged. This software package has several very useful utilities, an emergency unerase, a manual or automatic backup, and some utilities for creating batch files and managing directories and files under Windows.

Fastback Fastback Plus, (504) 291-7221, was one of the first *fast* backup software programs. This software is now past v3.0 and is easy to learn and use. It also compresses the data so that fewer floppy disks are needed.

Back-It 4 Back-It 4, from Gazelle Systems, (800) 233-0383, has been revised recently to v4.0. This new version uses very high density data compression—as much as 3 to 1. It is also very fast and uses a sophisticated error correction routine. Unlike some of the other systems, Back-It 4 will allow you to use differently formatted floppies at the same time.

PC Tools PC Tools comes bundled with a very good backup program. They have now decided to sell the backup program separately to anyone who doesn't want to buy the whole package.

XTree XTree is an excellent shell program for disk and file management, having several functions that make computing much easier. You can use it to easily copy files from one directory or disk to another. I often use it when I only want to backup a few files.

Q-DOS 3 Q-DOS 3, from Gazelle Systems, (801) 377-1288, is an excellent shell program similar to XTree. It can be used to select and copy files to another hard disk or floppies.

DOS XCOPY The XCOPY command is a part of MS-DOS versions higher than 3.2. Several switches can be used with XCOPY. (A switch is a /). For instance, XCOPY C:*.* A:/A will copy only those files that have their archive attribute set to "on" and doesn't reset the attribute flag. XCOPY C:*.* A:/M will copy the files and then reset the flag. Whenever a disk on A is full, you merely must insert a new floppy and hit F3 to repeat the last command. This will continue to copy all files that have not been backed up. XCOPY C:*.* A:/D:03-15-92 will copy only those files created after March 15, 1992. Several other very useful switches can be used with XCOPY.

Several more very good backup software packages are available. Check through the computer magazines for their ads and for reviews.

Tape

Several tape backup systems are out on the market. Tape backup is easy but can be relatively expensive—$400 to over $1000 for a drive unit and $5 to $20 for the tape cartridges. Most of them require the use of a controller that resembles the disk controller, so they will use one of your precious slots. Unless they are used externally, they will also require the use of one of the disk mounting areas. Because it's only used for backup, it will be idle most of the time.

If you have several computers in an office that must be backed up every day, you could possibly install a controller in each machine with an external connector. Then one external tape drive could be used to backup each of the computers at the end of the day. With this system, you would need a controller in each machine, but you would only have to buy one tape drive.

The tape systems are fairly well standardized today. One of the early problems with tape was that there were few standards for tape size, cartridges, reels, or format. Quite often a tape that was recorded on one tape machine, even from the same vendor, would not restore on another. That isn't much of a problem today. Most of today's tapes conform to the quarter-inch cartridge (QIC) standards, which cover several different formats and hardware interfaces. Several different tape capacities are available, from 50M up to over 500M.

The most common type drives use a quarter-inch tape similar to that used in audio cassettes. However, the data from a hard disk is much more critical than the cacophony of a rock band, so these tapes are manufactured to much stricter standards.

Also, half-inch drive systems are available for high-end use. They are much more expensive than the quarter-inch systems and will cost from $3000 to as much as $10,000 or more.

One of the big problems with software backup is that you must sit there and put in a new disk when one is full. A big plus for the tape is that it can be set up to run automatically. You don't need to worry about forgetting to backup or wasting the time doing it.

DAT

DAT stands for Digital Audio Tape. The audio record and tape industry has fought to keep the DAT systems out of the country, being worried that people will buy compact audio discs and then freely copy them onto DAT systems. Still, DATs are finally making their way into the country, although they are still very expensive.

Several companies are offering the DAT systems for backing up large computer hard disk systems. DAT systems offer storage capacities as high as 1.3 gigabytes on a very small cartridge.

The DAT systems use a helical scan type recording similar to that used for video recording. The DAT tapes are 4 millimeters wide, which is .156 inches and thus much smaller than the quarter-inch (.25") tapes. Yet they can store about twice as much data.

Here are just four of the many companies who are offering DAT systems:

Bi-Tech Enterprises	(516) 567-8155
Identica	(408) 727-2600
Tallgrass Technologies	(913) 492-6002
Tecmar	(216) 349-1009

Videotape

Another tape system uses video tape and a standard home videotape recorder to make backups. This system requires that an interface board be installed in the computer. From 60M to 120M of data can be stored on a standard videotape that costs from three to five dollars.

This type of system is ideal for the home user. Two leads from the interface easily connect to the VCR. After the backup is completed, the machine can be moved back to the living room for home entertainment. This low-cost system is also sophisticated enough to be used in large businesses and offices. The Alpha Micro VIDEOTRAX system has an option that will even do an automatic backup.

Besides being used for backup, these videotapes can be used to distribute large amounts of data. For instance, the contents of a hard disk could be copied and sent to another computer within the company across the room or across the country.

Alpha Micro has also demonstrated that one could broadcast software over the TV channels. A VCR can record the software as easily as it does an old movie, and then the software can be installed on any computer system equipped with an interface board.

Two companies are foremost in this type of backup:

AUTOFAX   (408) 438-6861
Alpha Micro VIDEOTRAX   (714) 641-6381

Their boards cost about $300 to $500.

Very high density disk drives

Several companies are now making extended high density 2.8M floppies and very high density 20M floppy disk drives. Insite peripherals, Brier Technology and several others have developed 3.5" floppy drives that can store 20M on a 3.5" floppy disk. The Bernoulli drive can put 44M on a 5.25" floppy disk.

Even if they cost a bit more, high density floppy drives can be more advantageous than tapes. The tape drives would only be used for backup, but a high density floppy would have much more utility, possibly even obviating the need for a hard disk.

For more details on these drives refer to Chapter 5.

Second hard disk

The easiest and fastest of all methods of backup is to have a second hard disk. You can easily install a second hard disk because most controllers have the capability of controlling a second hard disk. You would have to make sure that the second hard disk would work with your first disk, but that should be easy to determine. You would not need a large second disk—20M or 30M would be fine. With a second hard disk as a backup, you would not need a backup software package, which might cost $200 or more. You could probably buy a second hard disk for this amount.

An average hard disk will have an access speed of about 40 ms. Floppy disks operate at about 300 ms speed, which can seem like an eternity compared to the speed of even the slowest hard disk. Depending on the number of files, how fragmented the data is on the disk, and the access speed, a second hard disk can back up 20M in a matter of seconds. To back up 20M using even the fastest software will require 15 to 20 minutes, as well as a lot of disk swapping in and out of the drive. Depending on the type of disks and software that you use for backup, you might need 50–60 360K disks, 17 1.2M disks, or 14 1.44M disks. Some of the latest backup software makes extensive use of data compression, so fewer disks are needed.

Another problem with using software backup is that you can often not find a particular file. Most backup software stores the data in a system not the same as DOS files. Usually no directory similar to DOS is provided. Even the DOS BACKUP files show only a control number when you check the directory.

On one of my computers, I have a Conner 104M hard disk for my C drive and a Seagate 30M as a second hard drive. I backup my daily work to the 30M hard drive. Just to be on the safe side, I also use 1.2M and 1.44M floppies for my daily backups. I know; backing up both to a hard disk and to floppies is similar to a man who wears both a belt and suspenders. Still, I'd rather be safe than sorry. About once a month, I make a complete backup of my C drive on floppies.

External plug-in hard drives

Several companies are now manufacturing small 20-80M hard disks that operate off the parallel or serial connector. Many of them are battery-powered in order to be used with a lap top. They can also be used to backup data from a large desktop system or to transfer data from one system to another. Pacific Rim Systems, (415) 782-1013, makes an excellent 20M system. If you install the Stacker compression software on this 20M hard disk, it can store about 40M.

Hard cards

You can buy a hard disk on a card for $300 to $600, depending on capacity and company. The higher capacity disks are usually rather expensive, but you can also use compression software on these disks. You might find it worthwhile to install a card in an empty slot and dedicate it to backup.

If you don't have any empty slots, you might even consider just plugging in the card once a week or so to make a backup and then removing the card until you need it again. This would entail removing the cover from the machine each time, which could seem like a chore. However, I remove the cover to my computer so often that I only use one screw on it to provide grounding. Thus, I can remove and replace my cover in a very short time.

No matter what type of system or method used, if your data is worth anything at all, you should be backing it up. You might be one of the lucky ones that never need to use your backups, but it's much better to have backed up than to be sorry.

Chapter 8
Monitors

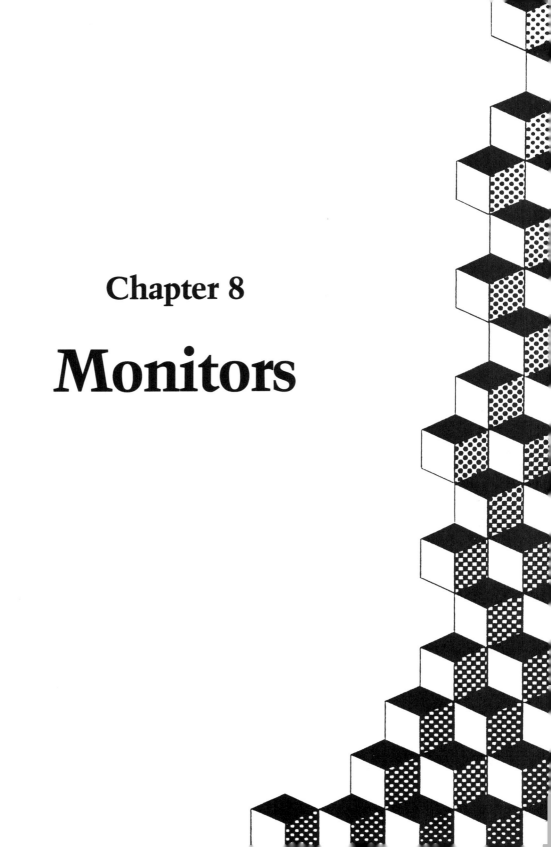

You can buy a monochrome monitor for as little as $65, but using a monochrome monitor reminds me of going to an art gallery and viewing all the paintings in black and white or of buying a black-and-white TV set. Life is too short to live it without color. Even if you do nothing but word processing, color makes the job much easier and more pleasant.

You can buy a fairly good color system for about $300. Alas, though, like so many other things in life, the better the color and the higher the resolution, the higher the cost. Also, the larger the screen or size of the monitor, the higher the cost. The design and development of the monitor electronics has been improved greatly, and new chipsets and VLSI integration have helped reduce the cost of manufacturing the electronics. Still, not much can be done to reduce the cost of manufacture of the main component—the cathode ray tube (CRT). A good color CRT requires a tremendous amount of labor-intensive, precision work. The larger the screen, the more costly it is to manufacture. Luckily, many companies manufacture monitors, so a great deal of competition helps keep the prices fairly reasonable.

What I recommend

When you start looking for a monitor, you will have a very wide choice of price, resolution, color, size, and shape. I would skimp on the rent and let the wife wear last year's styles before I would even consider a monochrome monitor. As a minimum, I would look for a 14" color VGA with at least a .31 dot pitch, which would cost around $300. A .28 dot pitch would give better resolution but would also cost more. If you can afford it, you might look for a 16" Super VGA (SVGA) that might cost from $600 to $1200 and up. Some of the brand name models such as NEC will cost a bit more. Luckily, the prices should be a bit lower by the time you read this; they come down every day.

If you are buying by mail or even at a store, try to get a copy of the manufacturer's specifications and study them. Look at the ads in magazines like the Computer Buying World, Computer Shopper, the Computer Monthly, PC Sources, and others to get an idea of the cost of a monitor. Be aware, though, that ads cost a lot of money, so a lot of good information is sometimes left out. Call the vendor and ask for a specification sheet. Monitors are usually long-lived. I have an early NEC Multiscan model about five years old. I have gone through several computers and hard disks, but this monitor is still going strong. (Of course, I've probably put a hex on it by saying this.)

The NEC Company was the first to develop a multiscan system. At that time, every one was running CGA and EGA digital systems. The analog VGA system was still just a gleam in IBM's eye. However, the NEC was designed so that it can be switched between digital or analog. I ran the monitor as a digital EGA for a couple of years; then, when the price of VGA adapters became reasonable, I bought one. I have never been sorry that I had to pay a little extra for this monitor. Almost all monitors manufactured today, even the monochrome, are analog VGA; digital

EGA and CGA are obsolete. If you are using your computer for something that doesn't require a lot of high resolution graphics, though, you might be able to find a good EGA bargain.

A checklist

We will go into some detail as to the mechanics of monitors later. If you're anxious to get started, though, here's a brief checklist for things you should look for. It's best if you can see several different models side-by-side. Even if you're going to buy through mail order, try to go to a computer swap meet or several different stores and compare various models. Compare the prices for equivalent models, and compare the mail order price—including shipping and handling. Here are some things that you might want to check:

Dot pitch Ask for a specification sheet and check for dot pitch. While it should be no less than 0.31mm for good resolution, 0.28mm or less is better.

Interlace Check for non-interlace if you expect to do a great deal of graphics work. Some monitors use interlacing in certain modes, so ask the vendor.

Scan rate Check scan rate, both vertical and horizontal. The less expensive ones have a fixed rate, while some have two or three fixed rates. The better ones use multiscanning.

Check bandwidth For a rough estimate, multiply the resolution times the vertical frequency. For instance, at a vertical frequency of 60Hz, the bandwidth would be $800 \times 600 \times 60Hz = 28.8MHz$, or a minimum of 30MHz. For a vertical frequency of 70Hz, it would be $800 \times 600 \times 70 = 33.6MHz$ or 35MHz minimum. For $768 \times 1024 \times 70$, it should be about 60MHz.

Check controls Some manufacturers put the controls in the back of the monitor, but it's better to have the controls accessible from the front or side. You'll have difficulty setting up the screen if you can't look at it at the same time, but once the monitor is set up, you shouldn't have to touch it again.

Check brightness Turn the brightness up and check the center of the screen and the outer edges; the intensity should be the same in the center and edges. Check for glare in different light settings.

Check with applications Check the focus, brightness, and contrast with text and with graphics. I have seen some monitors that displayed demo graphics programs beautifully but were terrible when displaying text in various colors.

Check for adapter and drivers Try to get an adapter that has as many drivers as possible with a minimum of 1M of VRAM, or at least one with 512K of VRAM and able to accept 512K more. You can add extra VRAM yourself.

Cables Make sure that you get the proper cables.

Tilt and swivel Even if you have to pay extra, get a monitor with a tilt-and-swivel base.

Monitor basics

I am going to present a few of the monitor specifications, terms, and acronyms so that you can make a more informed decision as to which monitor to buy.

In IBM language, a *monitor* is a display device. "Display" is probably a better term because the word "monitor" is from the Latin that means "to warn." Despite IBM, however, most people still call it a monitor.

Basically, a monitor works similarly to a television set. The face of a TV set or a monitor is the end of a Cathode Ray Tube (CRT), which is a vacuum tube and has many of the same elements that made up the old vacuum tubes used before the advent of the semiconductor age. The CRTs contain a filament that boils off a stream of electrons which have a potential of about 25,000 volts. The electrons are "shot" from an electron gun toward the front of the CRT, where they slam into the phosphor on the back side of the face and cause it to light up. Depending on the type of phosphor used, the dot (once lit) continues to glow for a period of time. The electron beam moves rapidly across the screen; but because the phosphor continues to glow for a while, we see the images that are created.

When we watch a movie, we are seeing a series of still photos flashed one after the other. Due to our persistence of vision, this flurry of pictures appears to be continuous motion. This same persistence of vision phenomenon allows us to see motion and images on our television and video screens.

In a magnetic field, a beam of electrons act very much like a piece of iron. Just like iron, a stream of electrons can be attracted or repelled by the polarity of a magnet. In a CRT, a beam of electrons must pass between a system of electro-magnets before it reaches the back side of the CRT face. In a basic system, one electromagnet would be on the left, one on the right, one at the top, and one at the bottom. Voltage through the electromagnets can be varied so that the beam of electrons is repelled by one side and attracted by the other, or pulled to the top or forced to the bottom. With this electromagnetic system, a stream of electrons can be bent and directed to any spot on the screen. This resembles holding a hose and directing a high pressure stream of water to an area: you could use the stream to write or draw lines or whatever.

Scan rates

When we look at the screen of a TV set or a monitor, we see a full screen only because of the persistence of vision and the type of phosphor used on the back of the screen. Actually, the beam of electrons start at the top left corner of the screen and then, under the influence of the electromagnets, is pulled across to the right-hand top corner. It lights up the pixels as it sweeps across, returns to the left-hand side, drops down one line, and then sweeps across again. On a TV set, this action is repeated so that 525 lines are written on the screen in about $1/60$ of a second. This one "write" to the screen is considered to be one *frame*, so 60 frames are written to the screen in one second.

The time that it takes to fill a screen with lines from top to bottom is the vertical scan rate. Some of the newer multiscan (multifrequency) monitors can have variable vertical scan rates from $1/40$ up to $1/100$ of a second to paint the screen from top to bottom. The horizontal scanning frequency of a standard TV set is 15.75kHz, which is also the frequency used by the CGA systems. The higher resolutions require higher frequencies: the horizontal frequency for the EGA is about 22kHz, while the VGA is 31.5kHz and the more advanced ones are higher.

The multiscan or multisync monitors can accept various frequencies sent to it from the adapter card, using horizontal signals from 15.5kHz up to 100kHz. Some of the older software was written specifically for CGA, EGA, or VGA; the multiscan can run any of them. Some of the low-cost monitors might accept only two or three fixed frequencies such as the 15.75kHz for CGA, the 22kHz for EGA, and the 31.5kHz for the VGA. Depending on your needs and bank account, this might be all you need. Still, the multiscanning is better if you can afford it.

Controlling the beam

The CRT has control grids (much like the old vacuum tubes did) for controlling the signal. A small signal voltage applied to the grid can control a very large voltage. The control grid—along with the electromagnetic system—controls the electron stream somewhat as if it were a pencil, causing it to make a large voltage output copy the small input signal. This amplified high voltage is then used to write a replica of the input signal on the screen. As the beam sweeps across the screen, if the input signal is tracing the outline of a figure, the control grid will turn the beam on to light up a dot of phosphor for the light areas. If the input signal is of a dark area, the beam is shut off so that a portion of the screen will be dark for that area of the image.

Monochrome

A monochrome monitor has a single electron beam gun and a single color phosphor. It writes directly on the phosphor and can provide very high resolution for text and graphics. It is even possible to get monochrome analog VGA, which can display in as many as 64 different shades. Large monochrome monitors might be ideal for some DeskTop Publishing systems (DTP) and even some Computer Aided Design (CAD) systems. These large monochrome monitors can be almost as expensive as the equivalent size in color.

Color

Color TVs and color monitors are much more complicated than monochrome systems. During the manufacture of the color monitors, three different phosphors— red, green, and blue—are deposited on the back of the screen, usually with a very small dot of each color being placed in a triangular shape. Three electron beam guns are used, one for each color. By lighting up the three different colored phosphors selectively, all the colors of the rainbow can be generated. The guns are called red, blue, and green (RGB), but they all emit the same electrons. They are called RGB simply because each gun is aimed so that it hits a red, green, or blue color dot on the back of the monitor screen. Of course, they are very accurately aimed so that they will converge or impinge only on their assigned color dot.

Resolution

If you look closely at a black and white photo in a newspaper, you can see that the photo is made up of small dots. A lot of dots appear in the darker areas, while fewer appear in the light areas. The text or image on a monitor or television screen is also made up of dots very similar to the newspaper photo.

If you want, you can easily see these dots with a magnifying glass. If you look closely, you can see spaces between the dots—these are much like the dots of a dot matrix printer. The more dots and tighter spacing, the better the resolution. A good high resolution monitor will have solid, sharply defined characters and images. An ideal resolution would look very much like a high quality photograph, but it will be some time before we reach the resolution of film.

Pixels

Resolution is also determined by the number of picture elements (i.e, *pixels*) that can be displayed. The following figures relate primarily to text, but the graphics resolution will be similar to the text.

CGA A standard Color Graphics Adapter (CGA) can display 640 × 200 pixels and 80 characters a line with 25 lines from top to bottom. If we divide 640 by 80, we find that one character will be 8 pixels wide. There can be 25 lines of characters, so 200/25 = 8 pixels high. The entire screen will have 640 × 200 = 128,000 pixels.

Most monitor adapters have text character generators built onto the board. When we send an A to the screen, the adapter goes to its library and sends the signal for the preformed A to the screen. Each character occupies a cell made up of the number of pixels depending on the resolution of the screen and the adapter. In the case of the CGA, if all the dots within a cell were lit up, there would be a solid block of dots 8 pixels or dots wide and 8 pixels high. When the A is placed in a cell, only the dots necessary to form an outline of an A will be lit up (which resembles dots being formed by the dot matrix printers when they print characters). A graphics adapter, along with the proper software, allows you to place lines, images, photos, normal and various text fonts, and almost anything else you can imagine on the screen.

EGA An Enhanced Graphics Adapter (EGA) system can display 640 × 350, or 640/80 = 8 pixels wide and 350/25 = 14 pixels high. The screen can display 640 × 350 = 224,000 total pixels.

VGA Super EGA (SEGA) and video graphics array (VGA) systems can display 640 × 480 = 307,200 total pixels, with each character being 8 pixels wide and 19 pixels high.

SVGA The Video Electronics Standards Association, (VESA) has chosen 800 × 600 to be the Super VGA (SVGA) standard, which is 800/80 = 10 wide and 600/25 = 24 high. Many of the newer systems are now capable of 1024 × 768, 1280 × 1024, 1664 × 1200, and more. With a resolution of 1664 × 1200, we would have 1,996,800 (i.e., almost 2 million) pixels that could be lit up. Obviously, we've come a long way from the 128,000 pixels possible with CGA.

XGA IBM has proposed that eXtra Graphics Array (XGA) be a new standard. It would have a resolution of 1024 × 768 and a vertical scan (refresh) rate of 70Hz.

CEG The Edsun Laboratories, (617) 647-9300, has developed a Continuous Edge Graphics (CEG) chip that can vastly increase the effective resolution of VGA monitors, smoothing out the edges of circles and lines in graphic images. Several companies are now incorporating this CEG chip on their adapter boards. You might have to spend a bit more for these adapters; but if you're doing a lot of graphics, they are well worth it. Eventually almost all adapters will use them or something equivalent.

The need for drivers and high resolution programs

You should be aware that even if you have a possible resolution of 1664×1200, your monitor will not display it unless you are using a driver or program that calls for it. If you are doing word processing or running an older program, your monitor will probably be displaying at 640×480 or even less.

Interlaced vs. non-interlaced

For CGA, the horizontal system will sweep the electron beam across the screen from top to bottom 200 times in $1/60$ of a second to make one frame, or 60 frames in one second. For VGA, it would sweep from top to bottom 480 times in $1/60$ of a second. For Super VGA, it would sweep 600 times; and for 1024×768 it would sweep 768 times in $1/60$ of a second.

As you can see, the higher the resolution you have, the more lines appear, the closer they are together, and the faster they have to be painted on the screen. The higher resolution also causes the electron beam to light up more pixels on each line as it sweeps across. The higher horizontal frequencies demand more precise and higher quality electronics, which of course requires higher costs to manufacture. To avoid this higher cost, IBM designed some of their VGA systems with an interlaced horizontal system. Instead of increasing the horizontal frequency, they merely painted every other line across the screen from top to bottom and then returned to the top and painted the skipped lines.

Theoretically, this sounds like a great idea, but it doesn't work too well in practicality because it causes a flicker. Some people who must work with this type of monitor for very long find it irritating. The flicker might not be readily apparent, but some people have complained of eyestrain, headaches, and fatigue after prolonged use of an interlaced monitor. If your head is turned slightly sideways, you might see from the corner of your eye the edges of the screen flickering. The flicker might be even more noticeable when you're doing intensive graphics.

Some companies make models that use interlacing in certain modes and non-interlacing in other modes. Most companies, though, don't advertise the fact that their monitors use interlacing. Still, the interlace models are usually a bit lower in price than the non-interlaced, and many of them also use the IBM standard 8514 chipset. You might have to ask the vendor what system is used.

Other companies besides IBM make interlaced monitors. If you get a chance, compare the interlaced and non-interlaced: you might not be able to tell the difference. If cost is a prime consideration, the interlaced is usually a bit less expensive and you find it a better deal.

The adapter that you buy should match your monitor: use an interlaced adapter with an interlaced monitor. An adapter that can send only interlaced signals might not work with a non-interlaced monitor. Some adapters are able to adjust and operate with both interlaced and non-interlaced monitors.

Dot pitch

The more color dots of phosphor and the closer together, the higher the possible resolution of a monitor. The distance between the dots per millimeter is called the *dot pitch*. A high resolution monitor might have a dot pitch of 0.31 millimeter (with 1 mm equaling 0.0394" and 0.31 mm equaling 0.0122"—about the thickness of an average business card). A typical medium resolution monitor might have a dot pitch of 0.39 mm, while one with very high resolution might have a dot pitch of 0.26 mm or even less.

The smaller the dot pitch, the more precise and difficult they are to manufacture. Some of the low-cost monitors have a dot pitch of 0.42 mm or even 0.52 mm. The 0.52 mm might be suitable for playing some games but would be difficult to use for any productive computing.

Bandwidth

The bandwidth of a monitor is the range of frequencies that its circuits can handle. The bandwidth also depends on the ability of the adapter to generate the proper frequencies. A multiscan monitor can accept horizontal frequencies from 15.75kHz up to about 40kHz and vertical frequencies from 40Hz up to about 90Hz.

To get a rough estimate of the bandwidth required, multiply the resolution pixels times the vertical scan or frame rate. For instance, a Super VGA, or VESA standard, monitor should have 800 × 600 × 60Hz = 28.8MHz. In addition, the systems require a certain amount of overhead, such as *retrace*—the time needed to move back to the left side of the screen, drop it down one line, and start a new line—so the bandwidth should be at least 30MHz. If the vertical scan rate is 90Hz, then it is 800 × 600 × 90 = 43.2MHz or at least 45MHz bandwidth. A very high resolution would require a bandwidth of 1600 × 1200 × 90 = 172.8MHz or about 180MHz, counting the overhead. Many of the very high resolution units are specified at 200MHz video bandwidth. Of course, the higher the bandwidth, the more costly and difficult to manufacture are the monitors.

Landscape vs. portrait

Most of our monitors are wider than they are tall; these are called *landscape* styles. Other monitors are taller than they are wide and are called *portrait* styles; they are often used for desktop publishing and other special applications.

Screen size

The stated screen size is measured diagonally the same way as the television screens, but there is usually a border on all four sides of the screen. The usable viewing area on my 13" NEC is about 9.75" wide and about 7.75" high. One reason is because the screen is markedly curved near the edges on all sides, which can cause distortion so that the areas are masked off and not used. This size is still big enough for most of the things that I do; but for some types of CAD work or desktop publishing, it would be helpful to have a bigger screen. Unfortunately, prices go up almost at a logarithmic rate for sizes above 14". A 14" might cost about $300, a 16" about $900, and a 19" or 20" might cost $2000 or more.

Controls

You might also check for available controls to adjust the brightness, contrast, and vertical/horizontal lines. Some manufacturers place them on the back or at some other difficult-to-reach area. You'll find the monitor easier to use if the controls are accessible from the front or side so that you can see how your adjusting them affects the screen.

Glare

If a monitor reflects too much light, it can act like a mirror and distract you from the work at hand. Some manufacturers have coated the screen with a silicon formulation that cuts down on the reflectance, while others have etched the screen for the same purpose. In addition, some screens are tinted to help cut down on glare. If possible, you should try the monitor under various lighting conditions. If you have a glare problem, several supply companies and mail order houses offer glare shields that cost from $20 up to $100.

Cleaning the screens

Because about 25,000 volts of electricity hit the back side of the monitor face, the screen develops a static attraction for dust. This dust can distort and make the screen difficult to read. Most manufacturers provide an instruction booklet that suggests how the screen should be cleaned. If you have a screen coated with silicon to reduce glare, you should not use any harsh cleansers on it. In most cases, plain water and a soft paper towel will do fine.

Tilt-and-swivel base

Most people sit their monitor on top of the computer. If you are short or tall, have a low or high chair or a non-standard desk, the monitor might not be at eye level. A tilt-and-swivel base can allow you to position the monitor to best suit you. Many monitors now come with this base. If yours does not have one, many specialty stores and mail order houses sell them for $15 to $40. Several supply and mail order houses also offer an adjustable arm that clamps to the desk, most having a small platform for the monitor to sit on. The arm can swing up and down and from side to side, which frees up a lot of desk space. These arms might cost from $50 up to $150.

Cables

Some of the monitors come without cables, but the vendors might sell them separately for $25 to $75 extra. Even those that have cables might not have the type of connectors that will fit your adapter. At this time, there is little or no standardization for cable and adapter connectors. Make sure that you get the proper cables to match your adapter and monitor.

Adapter basics

It won't do you much good to buy a high resolution monitor unless you buy a good adapter to drive it. You can't just plug a monitor into your computer and expect it to function. Just as a hard disk needs a controller, a monitor needs an adapter to interface with the computer. Also, like the hard disk manufacturers, many of the monitor manufacturers do not make adapter boards. Just as a hard disk can operate with controllers from several different manufacturers, most monitors can operate with adapters from several different manufacturers. The monitor and adapter should be fairly well matched. You could buy an adapter that has greater capabilities than the monitor can respond to, or your monitor might be capable of greater resolution than the adapter can supply.

Still, many 8-bit adapters are available (See Fig. 8-1). These are rather inexpensive and will be okay for general work; but if you expect to do any graphics or CAD type of work, you should definitely look for a 16-bit adapter such as that shown in Fig. 8-2. You should also look for one with at least 512K of video RAM; it would be better if it had 1M, but you can always add VRAM yourself.

The original IBM PC came with a green monochrome monitor with a monochrome display adapter (MDA) that could display text only. The Hercules Company immediately saw the folly of this limitation, so they developed the Hercules monographic adapter (HMGA) and set a new standard. It wasn't long before IBM and a lot of other companies were selling similar MGA cards that could display both graphics and text. These adapters provide a high resolution of 720 × 350 on monochrome monitors.

IBM then introduced their color monitor and color graphics adapter (CGA). CGA is a digital system that allows a mix of the red, green, and blue. The cables have four lines—one each for red, green, and blue, and one for intensity. This allows two different intensities for each color—on for bright or off for dim. Thus, there are four objects, each of which can be in either of two states, or two to the fourth power ($2^4 = 16$). Therefore CGA has a limit of only 16 colors. Also, the CGA monitors have very large spaces between the pixels, making the resolution and color terrible, comparable to the nine-pin dot matrix printer.

An enhanced graphics adapter (EGA) can drive a high resolution monitor to display 640 × 350 resolution, while Super EGA can display 640 × 480. The EGA system has six lines and allows each of the primary colors to be mixed together in any of four different intensities. Now there are 2^6 or 64 different colors that they can display, which makes the CGA system obsolete. I would not advise anyone to buy CGA.

Adapter basics **109**

8-1 An 8-bit VGA monitor board.

8-2 A 16-bit VGA monitor board.

Today the EGA system is also nearly obsolete, overrun by VGA. Still, just like with the obsolete 360K and 720K floppy drives, some vendors are still pushing the CGA and EGA systems. You might find a good buy on an EGA system. Depending on what you want to do with your computer and how much you want to spend, this might be all you need. However, for just a few dollars more than the cost of an EGA system, you can buy a much better VGA system.

Analog vs. digital

Up until the introduction of the PS/2 with VGA, most displays used the digital system. Unfortunately, the digital systems have severe limitations. The digital signals are of two states—either fully on or completely off. Also, the signals for color and intensity require separate lines in the cables. As pointed out above, it takes six lines for the EGA to be able to display 16 colors out of a palette of 64.

Contrastingly, the analog signals that drive the color guns are continuously variable voltages, and only a few lines are needed for the three primary colors. The intensity voltage for each color can then be varied almost infinitely to create as many as 256 colors out of a possible 262,144. The digital systems are sometimes called TTL for transistor-to-transistor logic. Some monitors that can handle both digital and analog might have a switch that says TTL for the digital mode.

Very high resolution graphics adapters

Many of the high resolution adapters have up to 1M or more of Video RAM (VRAM) memory on board. Some adapter boards are sold with only 512K of VRAM but probably have empty sockets for another 512K. You can buy VRAM and install it yourself. VRAM is about the same price as DRAM.

A single complex graphics drawing might require 1M of memory or more to store just one image. By having the memory on the adapter board, you don't have to go through the bus to the conventional RAM. Some adapter boards even have a separate plug-in daughter board for extra memory, and some have their own coprocessor on board (such as the Texas Instruments 34010 or the Hitachi HD63483). Depending on the resolution capabilities and the goodies that it has, a very high resolution adapter board might cost from $150 up to $3400.

The high resolution adapters are downward-compatible. If you run a program that was designed for CGA, it will display it in CGA format even though you have a very high resolution monitor. Still, it will look a lot better than it would on a CGA monitor.

Drivers

Be sure to ask your vendor about drivers for the adapter that you buy. Without the drivers, you might not be able to fully utilize the high resolution of your system. Most of the new software being developed today has built-in hooks that will allow it to take advantage of the high resolution goodies. The older software programs that were written before EGA and VGA were developed cannot normally take advantage of the higher resolution and extended graphics. Most manufacturers supply software drivers with their adapters for programs such as Windows, Lotus, AutoCAD, GEM, Ventura, and WordStar (plus others). Some vendors supply as many as two or three diskettes full of drivers. Depending on what software you intend to use, the drivers supplied with the adapter you purchase might be an important consideration.

Monitor and adapter sources

I have worked with many monitors in my lifetime, but there are hundreds that I have not had a chance to personally evaluate. I subscribe to Computer Buying World, PC Magazine, PC Week, Byte, PC Sources, InfoWorld, PC World, Computer Shopper, Computer Monthly, and about 50 other computer magazines. Most of these magazines have test labs and do extensive tests of products for their excellent reviews. Because I can't personally test all of these products, I rely heavily on their reviews. I can't possibly list all of the monitor and adapter vendors, so I suggest that you subscribe to at least the magazines I've just mentioned and those in the Appendix.

List price vs. street price

Note that the prices quoted from manufacturers in most magazine reviews are list prices. Often the "street price" of the products will be much less. For instance, here are some list prices from a recent magazine review and ad prices for the same units from a current Computer Shopper:

Model	List price	Ad price
NEC 2A 14"	$799	$394
NEC 3D 14"	$1049	$599
NEC 4D 16"	$1799	$1149
NEC 5D 20"	$3699	$2097

As you can see, there is a tremendous difference in the list price and the actual price you should pay. Also, the prices quoted here will be different by the time you read this. In this volatile market, the prices change almost daily. Call first if ordering from a magazine ad.

What to buy if you can afford it

The primary determining factor for choosing a monitor should be how you're going to use it and how much money you have to spend. If money is no object, buy a large 20″ analog monitor with super high resolution and a good VGA board to drive it (which will cost you about $2500). If money is important, you might consider a system such as the 20″ Sampo TriSync for about $1395. Their system uses three of the most common fixed frequencies.

You can call Sampo at (404) 449-6220. I've just bought one of their monitors, and for my type of work (and for the price), it seems to be a very good monitor. For databases, spreadsheets, accounting or for word processing, a monochrome monitor would probably be sufficient but not nearly as pleasant as a color monitor.

What should you buy? Buy the biggest and best multiscanning color Super VGA or XGA that you can afford.

Chapter 9

Memory

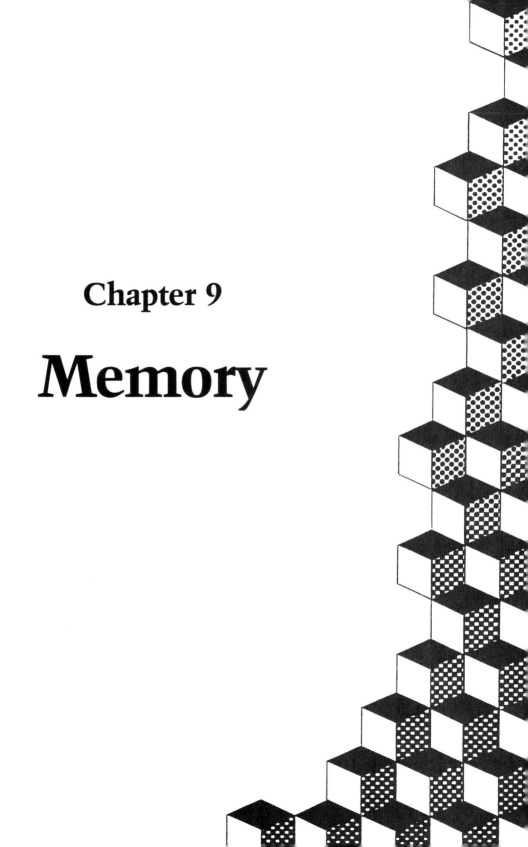

The computer uses two types of memory—Random Access Memory (RAM) and Read-Only Memory (ROM). ROM is unlike RAM in several respects. First of all, ROM is fixed/permanent memory: It compares to a book that can only be read. Conversely, RAM is like a blackboard that can be written on and then erased. The most common type of memory in use today is Dynamic RAM (DRAM), called dynamic because—due to its design—it must be constantly refreshed.

Programs and files are loaded into RAM while they are being edited, changed, or while a program is running. RAM is one of the most critical elements of the computer. If we open a file from a hard disk, the files and data are read from the disk and placed in RAM. When we load in a program—be it word processing, a spreadsheet, database, or whatever—we will be working in the system RAM. If we are writing, programming, or creating another program, we will also be working in RAM.

RAM allows us to randomly access the memory and then read and write to it. Here we can manipulate the data, do calculations, enter more data, edit, search databases, or do any of the thousands of things that software programs allow us to do. We can access and change the data in RAM very quickly. For most applications, the program is loaded into the standard 640K of RAM. Some applications will allow the use of up to 32M or more of RAM.

Besides the application programs that must be loaded into RAM, certain DOS system commands must be in RAM at all times, such as files like COMMAND.COM and other internal commands like COPY, CD, CLS, DATE, DEL, MD, PATH, TIME, TYPE, etc. These commands are always in RAM and are available immediately. The CONFIG.SYS file and any drivers that you might have for your system are also loaded into RAM.

Terminate and Stay Resident (TSR) files are loaded into RAM. These memory-resident programs, like SideKick Plus and others, can pop up any time you press a key. Portions of RAM can also be used for a very fast RAM disk, for buffers, and for print spooling.

All of these files contribute to the utility and functional ability of the computer and makes it easier to use. Unfortunately, they often take big bites out of your precious 640K bytes of RAM. You might have less than 400K left for running applications after loading DOS and a memory-resident program. To make matters worse, many programs are now so large that they won't fit in less than 600K of RAM.

DR DOS 5.0 from Digital Research and MS-DOS 5.0 from Microsoft can now load much of the operating system and TSRs in upper memory. Several other memory management programs have been developed to help alleviate this problem. One excellent program is DESQview from Quarterdeck Office Systems, (213) 392-9701.

An important difference in ROM and RAM is that RAM is *volatile*. In other words, it disappears if the machine is rebooted or if you exit one program and load another into memory. Also, if the power to the computer is interrupted, even for a brief instant, any data in RAM will be gone forever.

A brief explanation of memory

Computers operate on binary systems of 0s and 1s (i.e., off and on). A transistor can be turned off or on to represent the 0s and 1s. Two transistors can represent these four different combinations:

- both off.
- both on.
- #1 on and #2 off.
- #1 off and #2 on.

Correspondingly, a bank of four transistors can represent 16 different combinations; and with eight transistors, we now can have 256 different combinations. Eight transistors (called bits) make one byte. With them, you can represent each letter of the alphabet, each number, and each symbol of the extended American Standard Code for Information Interchange (ASCII). With eight lines, plus a ground, the eight transistors can be turned on or off to represent any single one of the 256 characters of the ASCII code.

Each byte of memory has a separate address. It would be laid out similar to a large hotel's "pigeon holes" for the room keys. It would have a row of pigeon holes for the rooms on each floor and would also be laid out so that the holes would be in columns. If the hotel had 100 rooms, you could have ten rows across and ten down. It would be very simple to find any one of the 100 keys by counting across and then down to the particular room number.

One megabyte of memory would require many more pigeon holes or cells. However, with just 20 address lines and one ground line, any individual byte can be quickly accessed.

DRAM

DRAM is relatively inexpensive, and large amounts can be packed onto small chips. As such, it's the memory of choice on almost all systems. DRAM is made up of small cells similar to small capacitors. A capacitor can be made from two conductive plates placed close together. When an electric charge is placed on the plates—a plus voltage on one plate and minus on the other—it will act like a battery. It will retain the charge for a certain amount of time, depending on the area of the conductors and certain other factors.

When it has a charge on it, a DRAM cell can represent a 1; when it is not charged, it could represent a 0. However, as soon as a cell is charged, it starts leaking and quickly becomes discharged. Thus, the computer must go back and recharge or refresh all of those cells that are supposed to be 1s. This refreshing might take up to 7% of a computer's time.

Refreshment and wait states

During the time that these cells are being refreshed, they cannot be accessed by your programs. After they have been accessed, they must be refreshed before they can be accessed again. If the CPU is operating at a very high frequency, it might have to sit and wait one cycle (one wait state) for the refresh cycle. This wait state might be only a millionth of a second or less, which might not seem like much time. However, if the computer is doing several million operations per second, it can add up.

It takes a finite amount of time to charge up the DRAM, and some DRAM can be charged up much faster than others. For instance, the DRAM chips needed for an XT at 4.77MHz might take as much as 200 nanoseconds (ns) or billionths of a second to be refreshed. A 386 running at 25MHz would need chips that could be refreshed in 70ns or less time. Of course, the faster chips cost more.

Interleaved memory

Some systems have been developed so that the memory is divided in half. One half of the memory would be refreshed on one cycle, and then the other half would be refreshed. If the CPU needed to access an address that was in the half already refreshed, it would be available immediately, thus reducing the amount of waiting by half or more. If the system uses interleaving, the memory must be installed in multiples of two (e.g., 2M, 4M, 8M, 16M or 32M). This could allow you to have a 20MHz 386SX system with 0 wait state using DRAM as slow as 80ns. Only those motherboards designed for interleaved memory can make use of it. When you buy your motherboard, ask the vendor whether it will accept interleave memory.

Cache memory

One way to make the computer operate faster is to install a small amount of very fast DRAM or SRAM in a cache near the CPU. The cache can be from 64K up to 128K or more. When running an application program, the CPU often loops in and out of certain areas and uses the same memory over and over. If this often-used memory is stored in the cache, it can be accessed by the CPU very quickly.

You can't arbitrarily add cache to any system; the motherboard usually has to be designed for it. If you are going to need a very fast system, you should look for a motherboard with a cache on it.

A cache system can speed up operations quite a lot. The computer could be slowed down considerably if it has to search the entire memory each time it has to fetch some data. The data used most frequently can be stored in the fast cache memory and speed might be increased by several magnitudes.

Cache memory should not be confused with disk caching. Often a program might need to access a hard disk while running. If a small disk cache is set up in RAM, the program will run much faster.

SRAM

Static RAM is made up of actual transistors. They can be turned on to represent 1s or left off to represent 0s and will stay that way until changed, removing the need to be refreshed. They are very fast and can operate at speeds of 25 ns or less but are much more expensive than DRAMs. A DRAM memory cell needs only one transistor and a small capacitor. Each SRAM cell requires four to six transistors and other components. Besides being more expensive, the SRAM chips are physically larger and require more space than the DRAM chips. So SRAM and DRAM chips are not interchangeable. Many laptops use SRAM because it can be kept alive with a small amount of current from a battery. Because DRAM must be constantly refreshed, a lot of circuitry and power is needed to keep them alive.

Flash memory

Intel has developed flash memory similar to Erasable Programmable Read Only Memory (EPROM). It is on small plug-in cards about the size of a credit card and is available on some small palm-size computers already such as the PSIONs. Flash memory is still rather expensive, but several foreign companies have begun developments along the same lines. A JEIDA standard has been proposed that would make the cards interchangeable among the various laptops.

Motherboard memory

The XT motherboard can accept 640K of memory on the motherboard. The 286 and 386SX can accept up to 16M but usually have sockets for only up to 8M. The 386DX and 486 might have sockets for 16M or 32M. Extra memory can be added to most systems with a plug-in memory board, but it would be limited to 16-bit memory unless it had a special slot. The 386DX and 486 are 32-bit systems and communicate with their on-board memory over a 32-bit bus, but they communicate with the plug-in slots over the standard 16-bit bus.

Dual In-line Package (DIP)

Most of the older motherboards used the DIP type of DRAMs, which are physically larger than the newer SIMMs and SIPs. The DIPs are practically obsolete now, but you might still find a few boards that use them. Some motherboards provide a combination of both DIP and SIP or SIMM sockets.

Single Inline Memory Module (SIMM)

Your computer motherboard will probably have sockets for SIMMs, which is an assembly of miniature 1M DRAM chips. There are usually nine chips on a small board that is plugged slant-wise into a special connector. They require a very small amount of board real estate. Figure 9-1 shows an older memory board with 4M of DIP memory installed. Below the board is a miniature SIMM package that has 1M. Figure 9-2 shows 4M of SIMM memory.

Single Inline Package (SIP)

Some motherboards might have SIP memory, which is similar to the SIMMs except that they have pins. See Fig. 9-3. Note that it takes 9 of the 1M chips to make 1 M. Each chip is actually only 128K, and the ninth chip does parity checking for errors. The 1000 marking on the chips indicates 1M. The 70 marking indicates a speed of 70 ns.

9-1 A plug-in memory board with DIP sockets and 4M of memory. Below it is a SIMM package with 1M of memory.

9-2 Four megabytes of SIMM memory in lower left corner.

At the present time, 1M chips are least expensive and most generally available. By the time you read this, 4M chips should be fairly reasonable. The 16M chips will be available soon.

If you plan to add extra memory, be sure that you get the kind and type for your machine. Also make sure that it is fast enough for your system.

9-3 Two SIP packages with 1M each.

How much memory do you need?

The amount of memory you need will depend primarily on how you intend to use your computer. When you run a program, data is read from the disk into RAM and operated on there. For word processing or small applications, you can get by with 640K. If you expect to use Windows, large databases, or spreadsheets, you should have at least 2M. Better yet would be a minimum of 4M.

Several companies make memory boards, and many of them have SIMM or SIP sockets that could accept up to 32M. One corollary concerning computer memory was stated rather well by Mr. Bob Howe, an analyst. He was quoted in Computer System News as saying "Memory is like Heroin—users will always use more." Someone else stated another fact that "The need for memory will expand to whatever amount is available." Having lots of memory is like having a car with a large engine. You might not need that extra power very often, but it sure feels great being able to call on it when you do need it or feel like using it.

Types of memory
Conventional memory
This is the 1M of memory that includes the 640K. The 384K of memory above 640K is reserved for video, ROM BIOS, and other functions.

Extended memory
Extended memory is memory that can be installed above 1M. If it weren't for the DOS 640K limitation, it would be a seamless continuation of memory. Windows

3.0 and several other software applications will let 286 and larger computers use this memory.

Expanded memory

Some large spreadsheets require an enormous amount of memory. A few years ago, in a rare instance of cooperation among corporations, Lotus, Intel, Microsoft, and some other large companies got together and devised a system and standard specification called LIM EMS 4.0. It allows a computer, even a PC or XT, to address up to 32M of expanded memory. The memory is divided into pages of 16K each. Expanded memory finds a 64K window not being used above the 640K of the 1M conventional memory, and pages of 16K expanded memory can be switched in and out of this window. LIM EMS also includes functions to allow multitasking so that several programs can be run simultaneously. The system can treat extra memory on the 286, 386, and 486 as extended memory with the proper software and drivers. (See Fig. 9-4.)

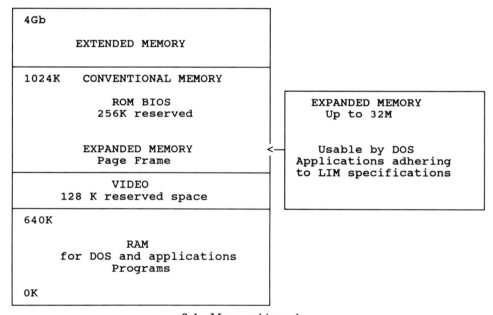

9-4 Memory hierarchy.

Memory modes

There are three different memory modes: real, standard or protected, and 386-enhanced.

Real mode

The real mode is the mode that most of us have been using until now. When any application is being processed, the program is loaded into RAM memory. The

CPU uses the RAM to process any data that is input. Computations, changes, or calculations are done in memory and then sent back to the disk, screen, printer, or other device. For most single user applications, this processing is done in the standard 640K or less of RAM.

Operating in the real mode doesn't cause much of a problem if you are running fairly small programs that can fit in the available RAM. However, if you are trying to update a spreadsheet that has 1M in it, you are in trouble. You might as well be trying to put a gallon of Jello in a quart-size bowl.

Breaking the 640K Barrier with LIM EMS The Lotus-Intel-Microsoft Expanded Memory Specification (LIM EMS) was designed to solve this problem. Expanded memory can be accessed, a small amount at a time, through memory above 640K. Expanded memory is like adding an extra room onto a building. When the building is full, you can store extra material in the added room. However, to use the extra material, you must go through a door and move a small amount in or out at any one time.

The LIM EMS scheme allows up to 32M of expanded memory to be added to a system. However, it can only be accessed through a small window above 640K. Only 64K of data can be moved in or out of the window at a time.

The LIM EMS system works on the XT as well as the larger AT type machines.

Windows 3.0 Because we now have Windows 3.0, there is not as much need for LIM EMS on the AT-type machines. Windows 3.0 lets you go beyond the 640K barrier, do multitasking, and take advantage of the virtual and protected modes. It works with a mouse and icons and can make your PC easier to use than a Macintosh. Better yet, Windows is relatively inexpensive.

DESQview DESQview will also let you break through the 640K barrier. It is an excellent program that will let you take advantage of the 386 virtual 8086 and 32-bit protected modes. It will let you run multiple DOS programs simultaneously, switch between them, run programs in the background, and transfer data between them. Also, DESQview is very inexpensive. For more details, call (213) 392-9701.

Standard or protected mode

The XT can only address 1M: that is, 2 to the 20th power. This 1M is 640K of lower memory and 384K of upper memory reserved for the ROM BIOS and video. The ROM BIOS and video usually does not require all of the reserved 385K, so portions of it can be used to access expanded memory.

The 286, the 386, and even the 486 is also limited to 640K in the real mode. With the proper software, however, the 286 can address 16M and the 386 can address 4 gigabytes. With Windows 3.0 and the proper application software, you can use extended memory on a 286, 386SX, 386, or 486.

In the protected mode with the proper software applications, you can also load two or more programs into memory and process both of them at the same time. If you tried this in the real mode, the data from both programs would be all mixed up together. Think of the data from both programs as just 0s and 1s. If you mix them together, it would be like mixing a gallon of hot water with a gallon of cold water.

However, the CPUs of the 286, 386, and 486 have a built-in system that can put a "wall" around a megabyte of RAM and let it work just as if it were a separate 8086 computer. Essentially, you can have several 8086 computers working on different applications all at the same time.

The 386 Enhanced mode

With the proper software applications, the 386SX, the 386DX, and the 486 can use all of the extended memory available. It will also set aside a portion of your hard disk and use it as virtual memory, which would allow programs as large as 32M or more to be processed all at once. The 286 cannot operate in this mode but otherwise can do just about everything the 386SX can do.

Chapter 10
Input devices

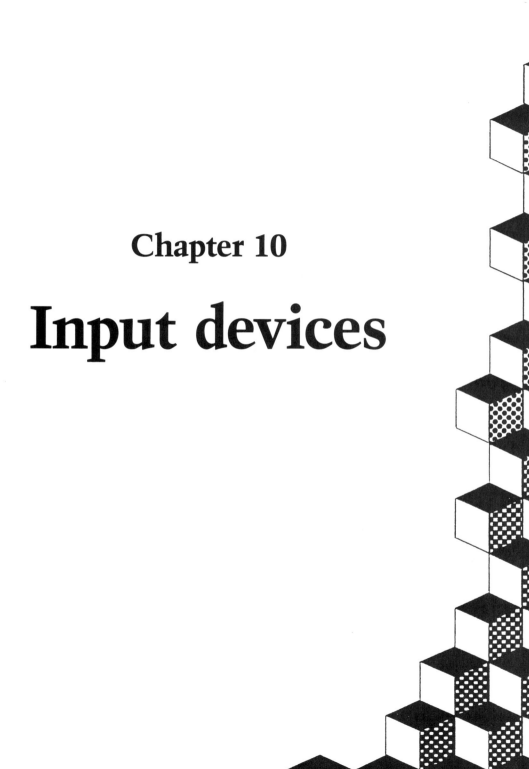

Before you can do anything with a computer, you must input data to it. You can input data in several ways, such as from a disk, by modem, by a mouse, by scanner, by bar code readers, by voice data input, by FAX, or on-line from a mainframe or a network. By far the most common way to get data into the computer, however, is by way of the keyboard. For most common applications, it is impossible to operate the computer without a keyboard.

The keyboard is your most personal connection with your computer. If you do a lot of typing, you must get a keyboard that suits you. Not all keyboards are the same; some have a light mushy touch, some are heavy, some have noisy keys, and others are silent with very little feedback. Personally, I have a very heavy hand. I bought a computer once that had a keyboard with very soft keys; if I just barely touched a key, it would take off. I finally took the keyboard to a swap meet and sold it for about half of what it cost me. Then I went around to all the booths at the show and tried several keyboards until I found one that had a touch that I liked.

That keyboard lasted for quite some time. One night, however, I decided to do some more work before going to bed. I had poured myself a glass of wine before bed and, without thinking, set the wine glass on the desk near the keyboard. Well, that was a mistake (you can easily guess what happened). I'll tell you one thing: keyboards can't hold their wine. I took the keyboard apart and cleaned it, but it never worked again.

The Quality Computer Products Company, (800) 752-1745, says that their QCP 101 keyboard is spill-proof. It has a list price of $42. I should either buy one of those keyboards or give up drinking wine near my computer.

Keyboard covers

Special plastic covers have been made that can protect against spills, dust, or other harsh environments. Sometimes a cover is absolutely essential. Most of the covers are made from soft plastic molded to fit over the keys; they are pliable but still slow down any serious typist.

Several companies make these covers. One company is CompuCover, (800) 874-6391, while another is Tech-Cessories at (800) 637-0909. Well over 400 different keyboards are used in the US alone; and if you count the foreign type of keyboards, you'll probably find over 4000 different types worldwide. These companies claim that they can provide a cover for most of them, with the average one costing about $25. This seems a bit expensive for just a bit of shrink-type plastic. Of course, one reason they are so expensive is because of the many different keyboards.

A need for standards

Typewriter keyboards are fairly standard. The alphabet only has 26 letters in it and just a few symbols, so most QWERTY typewriters have about 50 keys. Still, I have had several computers over the last few years, and every one of them have had a different keyboard. The main typewriter characters aren't changed or moved very often; but some of the very important control keys like the Esc, the Ctrl, the PrtScr, the \, the function keys, and several others are moved all over the keyboard. IBM can be blamed for most of the changes.

The original IBM keyboard had the very important and often-used Esc key just to the left of the 1 key in the numeric row. The 84-key keyboard moved the Esc key over to the top row of the key pad. The tilde and grave key was moved to the original Esc position to the left side of the 1. The IBM 101-key keyboard moved the Esc back to its original position.

For some unknown reason, IBM also decided to move the function keys to the top of the keyboard above the numeric keys. This is quite frustrating for WordStar users because the Ctrl key and the function keys are used quite often. The original position made them much easier to access.

I mentioned earlier that the US has well over 400 different keyboards. Many people make their living by typing on a keyboard. Some of the large companies have systems that count the number of keystrokes that an employee makes during a shift. If the employee fails to make a certain number of keystrokes, then that person can be fired. Can you imagine the problems if the person has to frequently learn a new keyboard? I am not a very good typist in the first place. If I have to use a new keyboard, I have great difficulty. Thus, I believe that there should be some sort of standard.

The 101-key keyboards are 20" long and take up about 30 percent more desk space than the 18" 84-key keyboards. If you have a large desk, that might not be important. One of my desks has a section in the middle lower than the rest of the desk. This makes the keyboard just the right height for comfortable typing. It is great with an 84-key keyboard, but there isn't enough room for a 101-key keyboard in this space. Figure 10-1 shows three keyboards. Notice that they're all different.

10-1 Three keyboards, all with different key positions.

Several companies have taken note of the complaints about wasted desk space and are producing a 101-key keyboard the same size as the 84-key one. Many of them also offer an option to change the Ctrl key back to where it should be—by the A.

Model switch

We should note that the PC, XT, AT, 80286, 80386, and 80486 keyboards all have the same connectors. Any keyboard will plug into any one of those machines, but the PC and XT keyboards have different electronics and scan frequencies. An older PC or XT keyboard can be plugged into an 80286 or 80386 machine but won't operate.

Some keyboards have a small switch on the back side that allows them to be switched so that they can be used on a PC or XT or on the AT type 286, 386, or 486 machines. Some of the newer keyboards can electronically sense the type of computer it is attached to and automatically switch to that type.

How a keyboard works

The keyboard is actually a computer in itself, having a small microprocessor with its own ROM. The computerized electronics of the keyboard eliminates the bounce of the keys, can determine when you hold a key down for repeat, can store up to 20 or more keystrokes, and can determine which key was pressed first if you press two at a time. The newer microprocessors for the AT type machines are more complex and sophisticated than the early PC types.

Each time a key is pressed, a unique signal is sent to the BIOS. This signal is made up of a direct current voltage turned on and off a certain number of times within a definite time frame. Each time a 5-volt line is turned on for a certain amount of time, it can represent a 1; and when it is off for a certain amount of time, it can represent a 0. In the ASCII code, if letter A is pressed, the binary code for 65 will be generated—1 0 0 0 0 0 1.

Special keys

The main part of the keyboard is very similar to a typewriter layout. However, I'll describe some of the extra keys.

You can press two or more special keys at the same time for some functions. Pressing a key such as the Shift, Ctrl, Alt, and function keys, in conjunction with other keys, can give us a very large number of virtual keys.

Function keys

The function keys are multi-purpose keys. What they do is usually dependent on the software you are using at the time. Many software programs use the function keys to accomplish a goal with a minimum of keystrokes. Quite often, the function of the keys will be displayed somewhere on the screen.

On the older keyboards, the function keys were located to the left of the standard keys. The older keyboards had 10 function keys, while the 101 keyboards have 12.

Macros for the function keys are already set up in many software programs. I am using WordStar to type this. WordStar has an installation menu that allows users to easily program the function keys any way they want. Ten functions are available by using each function key alone, such as on-line help, underline, boldface, delete a line, delete a word, etc. Thirty more functions are available by using the function keys with the Ctrl, Shift, and the Alt keys.

DOS and function keys

DOS provides a number of macros or shortcuts by using the function keys:

F1—Redisplay Each Character If you have entered a DOS command, pressing F1 will redisplay the command character by character each time the key is pressed.

F2—Change Part of Last Command If you have entered a DOS command and you want to change part of it, you can enter a letter of the command and all of the command up to that letter will be displayed. You can then change the command from that point onward. This can save a few keystrokes.

F3—Redisplay Last Command F3 will redisplay the entire command that was previously entered. For instance, you can enter the command COPY A: B: and when it has finished, if you want to make a second copy, just press F3 and the command will come up again. All you have to do then is press Enter.

F4—Use Part of Last Command If a command has been entered and you want to reuse the last portion of it, just press a letter of the command and the portion from that letter to the end of the command will be displayed.

F6—End of File Character When you create a file, such as a .BAT file, DOS has to know when the file ends. F6 puts the ^Z to mark the end of your files.

Numbers

The keyboard has two sets of numbers. The row across the top is very similar to the keys on a standard typewriter and also has the standard symbols available when the shift and a number key are pressed.

The second set of numbers are arranged in a configuration similar to that of a calculator keypad, which makes it very easy to input large amounts of numeric data. The numbers on this set of keys are active only when the Num Lock is on. When the Num Lock is off, the special keys move the cursor about the keyboard.

Arrow keys

The Down arrow (on the 2 key) moves the cursor down one line, while the Up arrow (on the 8 key) moves it up one line. The Left arrow (on the 4 key) moves the cursor to the left, while the Right arrow (on the 6 key) moves it to the right. The 101 key keyboard added a second set of arrow keys between the main keys and the number pad.

Home and End keys

The Home (on the 7 key) moves the cursor to the upper left hand corner of the screen, while the End (on the 1 key) moves it to the bottom of the screen or to the end of the text.

PgDn and PgUp (Page Down and Page Up)

The PgDn (on the 3 key) causes the next page to be displayed on the screen, while the PgUp causes the computer to page backwards through the file.

Ins (Insert)

The Ins (on the 0 key) allows text to be inserted without writing over the present text. The text will continue to move to the right to allow text to be entered as long as Insert is on. This key works like a toggle switch, turning on or off each time it is pressed. With the Insert off, anything typed will replace any character at the cursor.

Del (Delete)

The Del (on the period or decimal key) will delete the character at the cursor. All characters to the right will move to the left to fill in the gap.

The 101 keyboards have additional and separate PgUp, PgDn, Ins, Del, Home and End keys between the main keys and the numeric pad.

Esc (Escape)

Most programs utilize this key in some way, to leave a program, to erase a line, or many other functions.

PrtScr (Print Screen)

By pressing the PrtScr key, anything on the screen will be sent to the printer and printed out.

* (asterisk)

The * key has different meanings depending on the program or mode that you're in. For most calculating programs, it represents times or multiplication. For instance, 4*15 would mean 4 times 15. In DOS, the * can be used as a wildcard for copying or manipulating files. For instance, if you had several files that had the extension .BAK and you wanted to erase all of them, you could give the single command DEL *.BAK.

Scroll Lock

The Scroll Lock has different functions in different programs. Some software programs use it to disable the cursor control keys.

Break

When the Ctrl key is pressed along with the Break key, the computer will usually interrupt what it is doing. Pressing Ctrl-C does about the same thing.

On some keyboards, the Break is on the same key with Scroll Lock.

Backspace

The backspace key is a left-pointing arrow at the far right on the standard numeral key row. It moves the cursor backwards one character at a time. In most programs, it erases the character to the left of it as it moves backwards.

Return or Enter

Return or Enter used to tell the computer that you have finished a particular line or entry. In many word processors, the Return or Enter key is used only at the end of a paragraph. When the cursor reaches the end of your right margin, most word processors automatically wrap around and send the cursor to the left for the next line.

Shift

The keyboard has two Shift keys operating in the same manner as those on a typewriter. They are used to print the uppercase capital and the standard symbols on the numeric keys.

CapsLock

This key is similar to the Shift Lock key on a typewriter, except that it affects the letters of the alphabet only. If you want to type in a $ sign for instance, you must use the shift key whether or not the CapsLock is on. When the CapsLock is on, you can also use the Shift keys to type a lowercase letter of the alphabet.

Ctrl (Control)

The Ctrl key is used in conjunction with several other keys for a variety of purposes. If you ask to view a long directory with the DIR command, it might scroll up the screen very quickly. You can use the Ctrl plus S (Ctrl-S) to stop it and then press any key to start it again. Ctrl plus C (Ctrl-C) will abort the directory and return you to the prompt sign of whatever directory or file you were in.

Tab

The Tab key has a left- and a right-pointing arrow with a bar at the end of each arrow tip. The bottom arrow that points to the right works just like the Tab key on a typewriter. The top arrow that points left will move the cursor backwards to tab stops when the shift key is used with it. Not all programs allow the use of the backward tabs.

Alt (Alternate)

The Alt key can have a variety of functions depending on the particular software program, although it's most often used with Ctrl and Del to reboot or reset the computer system.

For example, sometimes the computer will hang up. You might have given the computer a command to go off and perform a certain task, which it will then try to do. If part of the program is missing or for some reason the task cannot be performed, the computer might continue to try to run the software and ignore any pleas from you to stop. Often, you have no other recourse but to use Ctrl-Alt-Del to reset the system. This clears the memory, and anything that has not been saved to disk will be erased. Sometimes even this "warm boot" will not clear the computer. In this case, you must turn off the power and turn it back on again to clear the memory (this procedure being called a "cold boot").

Other special key functions

Here are some other special key functions:

- \ Backslash. It's used by DOS to denote a subdirectory. To change from one subdirectory to another, you must type CD \ and then the directory name. If no name is given, you will be returned to the root directory. The backslash should not be confused with the slash. If you use the slash where the backslash should be used, you will get an error message.
- / Slash, virgule, shilling, or solidus. In calculations, it represents division.
- < Less than.
- \> Greater than.
- ^ Caret. It symbolizes exponents.
- + Plus. It represents addition.
- − Minus. It represents subtraction.
- = Equals.

Reprogramming key functions

The keys can be changed by various software programs to represent almost anything you want. One thing that makes learning computers so difficult is that every software program uses the special keys in a different way. You might learn all the special keys that WordStar uses, but if you want to use a word processor such as WordPerfect or Microsoft Word, you will have to learn the special commands and keys that they use.

Keyboard sources

Keyboard preference is strictly a matter of individual taste. The Key Tronic Company makes some excellent keyboards. They can even let you change the little springs under the keys to a different tension. The standard is 2 ounces, but you can configure the key tension to whatever you like. You can install 1, 1.5, 2, 2.5, or 3 ounce springs for an extra $15. Key Tronic also lets you exchange the positions of the CapsLock and Ctrl keys. Their keyboards have several other functions clearly described in their large manual, the most detailed of any company.

Hundreds of clone makers offer very good keyboards for $35 to $90. Look through any computer magazine. If at all possible, try the keyboards out and compare. If you are buying a system through the mails, ask about the keyboard options.

Specialized keyboards

Several companies have developed specialized keyboards. I have listed only a few of them here.

Quite often I have the need to do some minor calculations, for which the computer is great. Several programs such as SideKick, Windows, and WordStar have built-in calculators. However, most of these programs require that the computer be on and using a file. A keyboard available from the Shamrock Company, (800) 722-2898, and also from Jameco, (415) 592-8097, has a built-in solar powered calculator where the number pad is located. The calculator can be used whether or not the computer is on.

The Focus Electronic Corp., (818) 820-0416, has a series of specialized keyboards that have built-in calculators, function keys in both locations, extra * and \ keys, and several other goodies. Their FK-50001 keyboard has eight cursor arrow keys. With these keys, the cursor can be moved right or left, up or down, and diagonally up or down from any of the four corners of the screen. The speed of the cursor movement can be varied by using the twelve function keys. These eight cursor keys will do just about everything that a mouse can do. The Focus keyboards are now very popular. A recent Computer Shopper magazine had ads from 18 different companies who were selling Focus keyboards. The FK-5001 was advertised at prices from $80 to $99 by the various dealers.

The Datacomp DFK 2010 is very similar to the FK-5001. It has the function keys at the top and left side and also lets you switch the Ctrl key back to where it is supposed to be. It also has diagonal arrow cursor keys like the FK-5001. The discount price is $50.

Just as IBM set the standard for the PC, Key Tronic of Spokane, (509) 928-8000, has been the leader in keyboard design. Most of the clone keyboards are copied from the Key Tronic designs. Besides the standard keyboards, they have developed a large number of specialized ones. Instead of a key pad, one has a touch pad that can operate in several different modes. One mode lets it act like a cursor pad; by using your finger or a stylus, the cursor can be moved much the same as with a mouse. It comes with templates for several popular programs such as WordStar, WordPerfect, DOS, and Lotus 1-2-3.

Another Key Tronic model has a bar code reader attached to it, which can be extremely handy if you have a small business that uses bar codes. This keyboard would be ideal for a computer in a point of sale (POS) system.

Mouse systems

One of the biggest reasons for the success of the Macintosh is its ease of use. With a mouse and icons, you only have to point and click and not learn a bunch of commands and rules. Thus, a person who knows nothing about computers can become productive in a very short time.

The people in the DOS world finally took note of this and began developing Windows and other programs for the IBM and compatibles that take advantage of the mouse. Most of these software programs can be used without a mouse but operate much faster and better with a mouse. To be productive, a mouse is essential for programs such as Windows 3.0, CAD programs, paint and graphics programs, and many others.

You can't just plug in a mouse and start using it. The software, whether Windows, WordStar, or a CAD program, must recognize and interface with the mouse. Now mouse companies develop software drivers that allow their mouse to operate with various programs, with these drivers usually being supplied on a diskette. The Microsoft Mouse is the closest to a standard, so most other companies emulate the Microsoft driver.

Mouse types

Mouse types have not been standardized yet. Thus, different types of mice sometimes operate on different principles.

For example, some mice use optics with a LED that shines on a reflective grid. As the mouse is moved across the grid, the reflected light is picked up by a detector and sent to the computer to move the cursor.

For a design that demands very close tolerances, the spacings of the grid for an optical mouse might not provide sufficient resolution. You might be better off in this case with a high resolution mouse that utilizes a ball. Some of the less expensive mice have a resolution capability of only 100 to 200 dots per inch (DPI), although Logitech and other companies have developed mice with resolution of over 300 DPI.

The ball-type mouse has a small round rubber ball on the underside that contacts the desktop. As the mouse is moved, the ball turns. Inside the mouse, two flywheels contact the ball—one for horizontal and one for vertical movements. You don't need a grid for the ball mice, but you do need about a square foot of clear desk space to move the mouse about. The ball picks up dirt, so you should clean it often.

The IMCS Company, (805) 239-8976, has put a mouse in a pen-like configuration. Their Mouse-Pen has a barrel about 6" long and about a 1/2" square. The barrel has two buttons and the foot of the pen has a small ball that functions exactly like a mouse. The pen can be moved just as if you were writing.

Mouse interfaces

Most of the mice require a voltage, usually 5 volts. Some come with a small plug-in transformer that should be plugged into your power strip, while others let you insert an adapter between the keyboard cable connector and the motherboard connector.

Some mice require the use of one of your serial ports for their input to the computer, which might cause a problem if you already have a serial printer using COM1 and a modem on COM2. Some of the motherboards have built-in mouse ports, and some of these ports have mouse connectors for the PS/2 type mouse. If you own a standard mouse, you will need an adapter to connect it.

Microsoft, Logitech and several other mouse companies have developed a Bus mouse that interfaces directly with the bus and does not require the use of one of your COM ports. However, the systems come with a board that requires the use of one of your slots. Figure 10-2 shows the Logitech bus mouse with its board and some of the software that comes with it.

10-2 A Logitech bus Mouse, its plug-in board, and some of the software that came with it.

Mouse cost

You can buy a fairly good mouse for $50 to $100. One reason for the seemingly high price is that some companies include options of software packages and other goodies with their products. Another factor that pushes up the cost is the mouse's resolution. The higher the resolution, the higher the cost.

Many companies manufacture mouse systems. Check the ads in the computer magazines listed in Appendix A.

Trackballs

A trackball is a mouse that has been turned upside down. Like the mouse, the trackball must have a voltage from a transformer or other source, and it also requires a serial port or a slot if it is of the bus type.

Instead of moving the mouse to move the ball, you simply move the ball itself. The trackballs are larger than the ball in a mouse, so the possible resolution is better.

Trackballs usually do not require as much desk space as the ordinary mouse. If your desk is as cluttered as mine, then you definitely need a trackball.

Several companies manufacture trackballs, so look through the computer magazines for ads.

Keyboard/trackball combination

The Amtac Company (718) 392-1703, Chicony Corp. (714) 771-6151, and several other companies, have keyboards with a trackball built into the right-hand area. A person owning this keyboard thus has mouse benefits and capabilities without losing any desk real estate. The trackball is compatible with the standard Microsoft and Mouse Systems.

I bought a Chicony combination keyboard and trackball for $68. It isn't as easy to use as the mouse, but it's a bargain when you consider that a stand-alone trackball might cost $75 or more. Several other companies make combination trackball/keyboard systems, so check the computer magazines for ads.

Digitizers and graphics tablets

Graphics tablets and digitizers are similar to a flat drawing pad or drafting table. Most of them use some sort of pointing device that can translate movement into digitized output to the computer. Some are rather small, while some might be as large as a standard drafting table. The cost varies from as little as $150 up to over $1500. Most of the tablets have a very high resolution, are very accurate, and are intended for precision drawing.

Some of the tablets have programmable overlays and function keys, and some will work with mouse-like device, a pen light, or a pencil-like stylus. The tablets can be used for designing circuits, for CAD programs, for graphics designs, free-hand drawing, and even for text and data input. They are most commonly used with CAD type software, however.

Most of the tablets are serial devices, but some of them require their own interface board. Many of them are compatible with the Microsoft and Mouse Systems.

Scanners and optical character readers

Most large companies have mountains of memos, manuals, documents, and files that must be maintained, revised, and updated periodically. If a manual or document is in loose-leaf form, then only those pages that have changed must to be retyped. Quite often, though, a whole manual or document must be retyped and reissued.

OCR

Several companies now manufacture optical character readers (OCRs) that can scan a line of printed type, recognize each character, and input that character into a computer just as if it were typed in from a keyboard. Once the data is stored in the computer, a word processor can be used to revise or change the data and then print it out again.

If copies of the printed matter are also stored in a computer, they can be searched very quickly for any item. Many times I have spent hours going through printed manuals looking for certain items; if the data had been in a computer, I could have found the information in just minutes.

Optical character readers have been around for several years. When they first came out, they cost from $6,000 to more than $15,000. They were very limited in the character fonts that they could recognize and were not able to handle graphics at all. Vast improvements have been made in the last few years. Many full page scanners are now fairly inexpensive, starting at about $650.

Hand-held

Some hand-held scanners that are very limited might cost as little as $200. The more expensive models usually have the ability to recognize a large number of fonts and graphics.

I have a Caere Typist hand-held scanner, which can scan about 4.5" at a time. However, it can automatically line up and match the text on the next scan. You can even scan the text in vertically (i.e., as you would normally read down a column). You can also scan horizontally across the page, and the scanner will recognize the text and place it in the computer properly.

Some of the hand-held scanners can recognize both text and graphics, although not both at the same time. They must be set either for text or graphics.

One advantage of the hand-held scanner over the flat full-bed scanner is that an individual page of a book can be scanned. The page must be on a good solid surface and provide enough flat surface for the rollers on the back of the instrument. If the surface is not solid and flat, the scanner will have difficulty recognizing some text. Something such as a clipboard placed under the sheet to be scanned works fine.

The Houston Instruments Company specializes in manufacturing plotters. They have developed a scanning head for one of their plotters that can scan a large drawing, digitize the lines and symbols, and then input them to a computer. The drawing can then be changed and replotted very easily.

Many companies manufacture input devices. Look in any of the computer magazines listed in Chapter 15. You will see many ads for all types of keyboards, scanners, mice, and other input devices.

Some of the magazines such as Computer Buying World, PC Sources, Computer Monthly, and Computer Shopper have a separate product listing in the back pages, which is a great help. However, they list only those products that are advertised for that month in their magazine.

Of course, many manufacturers of good products can't afford the high cost of magazine ads. If you live in a larger city, you should know of some local computer stores. Also, most of these cities usually have computer swaps, where you can probably find some good products.

Chapter 11

Printers

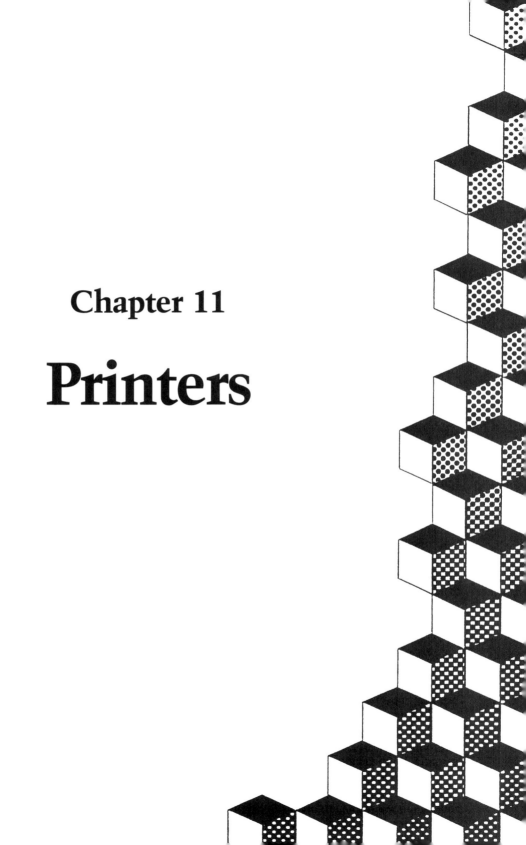

A computer system without a printer is about like ham without eggs; a printer is essential and necessary for almost all applications. Hundreds of different printers are available. Some dot matrix printers are relatively inexpensive, while some of the high-end lasers can be very expensive. A majority of the others lie somewhere between these two, and one of them should be able to fit your needs and budget.

Choosing a printer

The type of printer you need will depend primarily on how you plan to use your computer. If you can afford the price, a laser will be best for most applications. If at all possible, before you buy a printer, visit a computer store or a computer show and try it out. Get several spec sheets of printers in your price range and compare them. You should also look for reviews of printers in the computer magazines. PC Magazine has done an annual printer review every year since 1984. They have excellent test labs, a well-informed review staff, and their reviews and tests are comprehensive, thorough, and unbiased. You can order some of their back issues to check through their reviews.

Dot matrix printers

Dot matrix printers range greatly in price. The low-priced ones are not very fast, the print quality might be poor, and they might be limited in fonts and graphics capability. In contrast, some of the higher-priced dot matrixes can print in near-letter quality (NLQ) at a speed equivalent to some of the lasers. Many of them can also print different fonts and graphics. Some of them can even print fairly good color by using low-cost multicolor ribbons.

Many applications need only a dot matrix printer to accomplish a task. For example, wide continuous sheets are necessary for some spreadsheet printouts. Unfortunately, my LaserJet can't handle anything wider than 8.5". With the wide carriage on my Star dot matrix, the wide sheets are no problem.

Most of the dot matrix printers sold today are 24 pins, are fairly reasonable in price, and are sturdy and reliable. The 24-pin printer forms characters from two vertical rows of 12 pins in each row. Small electric solenoids surround each of the wire pins in the head. By pressing various pins as the head moves across the paper, any character can be formed. You can also print graphics. Some of the less expensive printers have a vertical row of only 7 or 9 pins, while the more expensive ones have 18 or 24. As the head moves in finite increments across the paper, solenoids push individual pins forward to form characters. At the top of the next page is a representation of the pins in a 7-pin print head and how it would form the letter A.

The numbers on the left represent the individual pins in the head before it starts moving across the paper. The first pin to be struck would be number 7, then number 6, then 5, 4, 3, 5 and 2, 1, 2 and 5, 3, 4, 5, 6, and then 7.

A 24-pin head would be similar to the 7-pin representation shown here, except that it would have two vertical rows, side by side, of 12 pins in each row. The pins

(Print head moves toward the right)

in one row would be slightly offset and lower than the pins in the other row. Because the pins are offset, they would overlap slightly and fill in the open gaps normally found in a 7- or 9-wire system.

Dot matrix speed

In draft mode, the speed can be from 200 characters per second (CPS) up to 600 or more. In near-letter quality (NLQ) mode, it might be from 60 up to 330 CPS. At 60 CPS, it would take a little over one minute to print a full page of 4000 characters. At 330 CPS, it would take less than 15 seconds or about 4 pages per minute (PPM).

Some high-end dot matrix line printers can print a whole line at once and are faster than most lasers. Some high-end machines might have two to four heads so that the line to be printed is divided up among the heads. The first head would print one fourth of the line, while the next head was printing the next fourth, and the other two were also printing the rest of the line.

Dot matrix color

Some dot matrix printers use a ribbon with three different stripes of colors—usually red, green and blue. Because the head can strike the various colors, all of the colors in the rainbow can be blended and printed. Of course, you need special software to accomplish this. Also, the printing is a bit slow; but if you want color to jazz up a presentation or for accent now and then, they're great. The color option will usually cost about $100 more than the standard price.

Advantages of dot matrix

Dot matrix printers can actually do some things that the laserjet can't. For instance, a dot matrix can have a wide carriage, while most lasers are limited to 8.5" by 11". Also, the dot matrix can use continuous sheets or forms, while the laser uses cut sheets, fed one at a time. In addition, a dot matrix can also print carbons and forms with multiple sheets.

Noise reduction

A dot matrix printer can be very noisy. Some enclosures are marketed that help to reduce the noise, but they're a bit expensive. I sat my printer on a 2" sheet of rubber foam that had been used as packing material, and the noise was considerably reduced.

Low cost

If you can get by with a dot matrix, you should be able to find some at very good prices. The low cost of the lasers are forcing the dot matrix people to lower their prices. In addition to lower prices, many dot matrix companies are also adding features such as more memory and several fonts in order to attract buyers.

Many vendors are still pushing the 9-pin dot matrix machines. A current issue of the Computer Shopper has an ad for a 9-pin Panasonic KXP NX 1000 for $119. A 9-pin Panasonic KX-P1180 with several built in features was advertised for $145, while a wide carriage 9-pin Panasonic KX-P1191 was advertised for $199. A 24-pin Panasonic KX- P1124 was advertised for $259, while a wide carriage 24-pin Panasonic KX-P1624 was advertised for $339.

You'll benefit from shopping wisely. I saw several other ads for these same printers with much higher prices. One ad listed the KX-P1124 for $295.95 and the KX-P1624 for $385.95.

For some heavy duty office work where a lot of multiple sheet forms are printed, a 9-pin dot matrix might be preferable to a 24-pin.

Print buffer

The computer can feed data to a printer much faster than it can print it. Thus, most printers have a buffer that can hold from 1K bytes up to 50K or more. The data is loaded into the buffer by the computer and is then fed to the printer as needed. With a large buffer, the computer can dump the data and then go about its business doing other things. With a small buffer and a long file, however, the computer will have to sit there and continually load the data into the buffer.

Daisy wheel printers

The daisy wheel has excellent letter quality. It has a wheel with all the letters of the alphabet on flexible "petals." If the letter A is pressed, an electric solenoid hammer hits the A as it spins by and presses it against a ribbon onto the paper.

Unfortunately, the daisy wheel printers are very slow, cannot print graphics, and are also quite noisy. Thus, now they are practically obsolete.

Ink jet

The ink jets are similar to the dot matrix in one respect. The dot matrix uses 9 to 24 pins to impact against a ribbon to create characters or graphics. Correspondingly, the ink jets use from 30 to 60 small spray nozzles to create characters or graphic images. Most of the ink jet machines print at a 300 dot per inch (DPI) resolution, the same as most lasers.

The HP DeskJet Plus is a small printer that has quality almost equal to that of a laser. It just uses a matrix of small ink jets instead of pins. As the head moves across the paper, the ink is sprayed from the jets to form the letters. It comes with Courier fonts but can use several more fonts available on plug-in cartridges. It has a speed of 1 to 2 PPM, is small enough to sit on a desk top, and is very quiet. I've seen it currently advertised for $549.

The ink jet wells are good for about 300 pages of text and must then be replaced or refilled. Luckily, this is relatively inexpensive and easy to do.

Cannon also manufactures a couple of printers based on the ink jet technology, but they call it the bubble jet. Their BJ-10e is a small portable model advertised for $309. Their desktop model BJ-130e is advertised for $647.

The Diconix division of Kodak also makes a small portable ink jet printer advertised for $341.00. These two portables are ideal for attaching to lap tops; you can have excellent letter quality print on the road.

Ink jet color

Hewlett-Packard has two models of color printers—the HP PaintJet for 8.5" × 11" and the HP PaintJet XL that can handle paper as large as 11" × 17". They can provide color by using ink cartridges with four different colored inks—black, cyan, yellow, and magenta. They offer a low-cost method of good quality color. A few laser-type color printers rely on a thermal wax transfer method to create color, but they are four to five times more expensive than an ink jet.

The color ink jet printers are ideal for creating colored transparencies for presentations, for graphs, and for schematic plotting and drawings.

Laser printers

Lasers have excellent print quality. They are a combination of the copy machine, computer, and laser technology. On the down side, they have lots of moving mechanical parts and are rather expensive.

Laser printers use synchronized, multifaceted mirrors and sophisticated optics to write the characters or images on a photosensitive rotating drum. The drum is similar to the ones used in reproduction machines. The laser beam is swept across the spinning drum and turned on and off to represent white and dark areas. As the drum is spinning, it writes one line across the drum, and then rapidly returns and writes another. It is quite similar to the electron beam that sweeps across the face of the monitor one line at a time.

The drum is sensitized by each point of light that hits it, and the sensitized areas act like an electromagnet. The drum rotates through the carbon toner, and the sensitized areas become covered with the toner. The paper is then pressed against the drum. The toner that was picked up by the drum leaves an imprint of the sensitized areas on the paper. The paper then is sent through a heating element where the toner is heated and fused to the paper.

Except for the writing to the drum, this is the same thing that happens in a copy machine. Instead of using a laser to sensitize the drum, a copying machine takes a photo of the image to be copied, and a photographic lens focuses the image onto the rotating drum.

Engine

The drum and its associated mechanical attachments is called an *engine*. Canon, a Japanese company, is one of the foremost makers of engines. They manufacture them for their own laser printers and copy machines, as well as for dozens of other companies such as Hewlett-Packard and Apple. Several other Japanese companies also manufacture laser engines.

The Hewlett-Packard LaserJet (see Fig. 11-1) was one of the first low-cost lasers. It was a fantastic success and became the *de facto* standard. There are now hundreds of laser printers on the market, most of them emulating the LaserJet standard. HP is the IBM of the laser world. Almost all of the lasers emulate the HP LaserJet series, even IBM's version.

11-1 My LaserJet III printer.

Low-cost laser printers

The competition has greatly benefited us consumers, driving down the prices and forcing many new improvements. Several new low-cost models have been introduced that print 4 to 6 pages per minute instead of the 8 to 10 pages of the original models. They are smaller than the originals and can easily sit on a desktop. Most have 512K of memory with an option to add more. The discount price for some of these models is now down to less than $800. The original 8 to 10 page models have dropped from about $3500 down to around $1000. If you can afford to wait a few seconds, the 4 to 6 page models will do almost everything the 8 to 10 pagers will do. The prices will drop even more as the competition increases and the economies of scale in the manufacturing process becomes greater.

An ad in the Computer Shopper offers a Panasonic KX-P4420 laser for $784. It can print 8 pages per minute (PPM), comes with 22 fonts, 512K memory, and several other goodies. The Hewlett-Packard LaserJet III has about the same features as this printer and is advertised for $1498 in the same issue of the Computer Shopper (which, you should note, is about twice the cost of the Panasonic). The HP LaserJet IIP, a 4 PPM printer, is advertised for $799.

For doing graphics, PC Magazine reported that the inexpensive 4 PPM units could print a page of graphics at about the same speed as the 8 PPM units. If you need a laser primarily for graphics, you can save quite a lot of money if you buy the 4 PPM units.

Extras for lasers

Don't be surprised if you go into a store to buy a laser printer that was advertised for $1000 and end up paying a lot more than that. The laser printer business is much like the automobile, computer, and most other businesses. I have seen laser printer ads for a very low price. Then, somewhere in the ad, in very small print, they say "w/o toner cartridge and cables." They might charge up to $150 for the toner cartridge and up to $50 for a $5 cable. Extra fonts, memory, special controller boards, and software will also cost extra. Some printers have small sheet bin feeders, and a large size might cost as much as $200 or more.

Memory

If you plan to do any graphics or desktop publishing (DTP), you must have at least 1M of memory in the machine. Of course, the more memory, the better. The laser memory chips are usually in SIMM packages. Not all lasers use the same configuration, so check before you buy. Several companies offer laser memories. Here are two:

    ASP (800) 445-6190
    Elite (800) 942-0018

Page description languages

If you plan to do any complex desktop publishing, you might need a page description language (PDL) of some kind. Text characters and graphics images are two different species of animals. Monitor controller boards usually have all of the alphabetical and numerical characters stored in ROM. When we press the letter A from the keyboard, it dives into the ROM chip, drags out the A, and displays it in a precise block of pixels wherever the cursor happens to be. These are called *bitmapped* characters. If you wanted to display an A that was twice as large, you would have to have a complete font set of that type in the computer.

With a good PDL, the printer can take one of the stored fonts and change it, or scale it, to any size you want. These are called *scalable fonts*. With a bitmapped font, you have one type face and one size. With scalable fonts, you might have one typeface with an infinite number of sizes. Most of the lasers printers will accept ROM cartridges that have as many as 35 or more fonts. You can print almost anything that you want with these fonts if your system can scale them.

Laser speed

Laser printers can print from four to over ten pages per minute depending on the model and what they are printing. Some very expensive high-end printers can print over 30 pages per minute.

A dot matrix printer is concerned with a single character at a time. The laser printers compose and then print a whole page at a time. With a PDL, many different fonts, sizes of type, and graphics can be printed. However, the laser must determine where every dot that makes up a character or image is to be placed on the paper before it is printed. This composition is done in memory before the page is printed. The more complex the page, the more memory required and the more time needed to compose the page. It might take several minutes to compose a complex graphics but, once composed, will print out very quickly.

A PDL controls and tells the laser where to place the dots on the sheet. Adobe's PostScript is the best known PDL. Several companies have developed their own PDLs. Of course, none of them are compatible with the others. This has caused a major problem for software developers because they must try to include drivers for each one of these PDLs. Several companies are attempting to clone PostScript, but they probably can't achieve 100% compatibility. Unless you need to move your files from a machine that does not have PostScript to one that does, you might not need to be compatible.

Hewlett-Packard includes their Printer Control Language 5 (PCL), a scalable font system, on their LaserJet IIIs.

PostScript printers

Printers sold with PostScript installed, such as the Apple LaserWriter, might cost as much as $1500 more than one without PostScript. Hewlett-Packard is offering a PostScript option for their LaserJet IID (the IID and IIID print duplex on both sides of the paper) for a list price of $995. The PostScript option for the IIP and III printers is $695, but the street prices should be less.

Pacific Data Products, (619) 552-0880, has a cartridge with built-in Bitstream fonts that is comparable to Adobe's PostScript.

PostScript on disk

Several software companies have developed PostScript software emulation. One of the better ones is LaserGo's GoScript, (619) 450-4600. QMS, (800) 631-2692, offers UltraScript; and the Custom Applications Company, (508) 667-8585, has Freedom of Press.

Resolution

Almost all of the lasers have a 300 × 300 dots per inch resolution (DPI), which is very good but not nearly as good as 1200 × 1200 dots per inch typeset used for standard publications. Several companies have developed systems to increase the number of dots to 600 × 600 DPI or more. LaserMaster, (612) 944-6099, has developed printer controllers that plug into a slot in the computer. They have several models that will increase resolution from 400 × 400 up to 1000 × 1000. At this time, of course, they are rather expensive. Eventually they will become more reasonable in price.

At 300 × 300 DPI, you can print 90,000 dots in one square inch. On an 8.5″ × 11″ page of paper, if we deduct a 1″ margin from the top, bottom, and both sides, then we would have 58.5 sq.in. × 90,000 dots = 5,265,000 possible dots.

Paper size

Most laser printers are limited to 8.5″ × 11″ paper (A size). The QMS PS-2200, (800) 631-2692, and the Unisys AP 9230, (215) 542-2240, can print 11″ × 17″ (B size) as well as the A size.

Maintenance

Most of the lasers use a toner cartridge that is good for 3000 to 5000 pages. The original cost of the cartridge is about $100. Several small companies are now refilling the spent cartridges for about $50 each.

Of course, there are other maintenance costs. Because these machines are very similar to the repro copy machines, they have a lot of moving parts that can wear out and jam up. Most of the larger companies give a mean time between failures (MTBF) of 30,000 up to 100,000 pages. Remember, though, that these are only average figures and not a guarantee.

Paper

Paper has many different types and weights, and almost any paper will work in your laser. If you use a cheap paper in your laser, however, it could leave lint inside the machine and cause problems in print quality. Generally speaking, any bond paper or a good paper made for copiers will work fine. Colored paper made for copiers will also work fine. As always, though, some companies are marking copy papers with the word "laser" and charging more for it.

Many of the laser printers are equipped with trays to print envelopes. Hewlett-Packard recommends envelopes with diagonal seams and gummed flaps. Make certain that the leading edge of the envelope has a sharp crease.

Address labels

The Avery Company, (818) 858-8387, has developed address labels that can withstand the heat of the fusing mechanism of the laser. Most office supply stores carry the labels.

Avery also developed an excellent software program that can be used to print out the labels. The program has a database where addresses, phone numbers, and other information can be stored. Any of the addresses can be searched for, sorted, edited, or printed out. The program can also read and import data and files from dBASE and WordPerfect, as well as import .PCX and .PCC graphics files.

Avery has both laser and dot matrix versions of the program and is ideal for anyone who does a lot of mailing.

Other specialty supplies can also be used with your laser. The Integraphix Company, (800) 421-2515, carries several different items that you might find useful. Call them for a catalog.

Color

A few color printers are available at a cost of $7000 and up at this time, although the prices should be a little less by the time you read this. These printers are often referred to as laser color printers, but they don't actually use the laser technology. They use a variety of thermal transfer technologies using a wax or rolls of plastic polymer. The wax or plastic is brought into contact with the paper, and then heat is applied. The melted wax or plastic material then adheres to the paper. Very precise points, up to 300 dots per inch, can be heated. By overlaying three or four colors, all of the colors of the rainbow can be created.

The cost of color prints ranges from about 5 cents a sheet for the Howtek Pixelmaster, which uses wax material similar to crayons, up to about 83 cents apiece for the large 11 × 17 inch sheets from the QMS ColorScript 30. A PC Magazine editor said that they save about $750,000.00 a year by using a color printer for corporate use rather than a photographic process. Another big plus is that the results from a color printer are available almost immediately. Any errors or corrections can be easily made.

Most of the color printers have PostScript or emulate it. The Tektronix Phaser CP can also use the Hewlett-Packard Graphics Language (HPGL) to emulate a plotter. These color printers can print out a page much faster than a plotter.

Several other color printers will be on the market soon. The large competition will then help drive the prices down.

Plotters

Plotters are devices that can draw almost any shape or design under the control of a computer. A plotter can have from one and up to eight or more different colored pens. Several different types of pens are made for various applications such as writing on different types of paper or on film or transparencies. Some pens are quite similar to ballpoint pens, while others have a fiber-type point. The points are usually made to a very close tolerance and can be very small so that the thickness of the lines can be controlled. The line thicknesses can be very critical in some precise design drawings.

The plotter arm can be directed to choose any one of the various pens. This arm is attached to a sliding rail and can be moved from one side of the paper to the other. A solenoid can lower the pen at any predetermined spot on the paper for it to write.

While the motor is moving the arm horizontally from side to side, a second motor moves the paper vertically up and down beneath the arm. This motor can move the paper to any predetermined spot and the pen can be lowered to write on that spot. The motors are controlled by pre-defined X,Y coordinates. They can move the pen and paper in very small increments so that almost any design can be traced out.

Values could be assigned of perhaps 1 to 1000 for the Y elements and the same values for the X or horizontal elements. The computer could then direct the plotter to move the pen to any point or coordinate on the sheet.

Plotters are ideal for such things as printing out circuit board designs, for architectural drawings, for making transparencies for overhead presentations, for graphs, for charts, and for many CAD/CAM drawings. All of these can be done in as many as eight or more colors.

Plotters come in several different sizes. Some desk top units are limited to only A and B sized plots, while other large floor-standing models can accept paper as wide as four feet and several feet long.

A desk model might cost as little as $200 and as much as $2000. A floor-standing large model might cost $4000 – $10,000. If you are doing very precise work (such as designing a transparency that will be photographed and used to make a circuit board), you will want one of the more accurate and more expensive machines.

Many good graphics programs are available that can use plotters. There are still several manufacturers of plotters, though, and no real standards. Just like the printers, each company has developed its own drivers. This is very frustrating for software developers who must try to include drivers in their programs for all of the various brands.

Hewlett-Packard has been one of the major plotter manufacturers, so many of the other manufacturers now emulate the HP drivers. Almost all of the software that requires plotters include a driver for HP. If you are in the market for a plotter, try to make sure that it can emulate the HP.

Houston Instruments is also a major manufacturer of plotters, and their models are somewhat less expensive than the Hewlett-Packard.

One disadvantage of plotters is that they are rather slow. Some software programs now allow laser printers to act as plotters. Of course, the printers are much faster than plotters but (except for the colored printers) are limited to black and white.

Plotter supplies

You should keep a good supply of plotter pens, special paper, film, and other supplies on hand. Plotter supplies are not as widely available as printer supplies. A very high-priced plotter might have to sit idle for some time if the supplies are not on hand. Most of the plotter vendors provide supplies for their equipment. One company that specializes in plotter pens, plotter media, accessories, and supplies is Plotpro, (800) 223-7568.

Installing a printer or plotter

Most IBM compatible computers allow for four ports—two serial and two parallel. No matter whether it is a plotter, dot matrix, daisy wheel, or laser printer, it will require one of these ports. If you have a 286 or 386 computer, these ports might be built into the motherboard. (See Chapter 10 and the discussion for installing modems).

If you have built-in ports, you will still need a short cable from the motherboard to the outside. You will then need a longer cable to your printer. If you don't have built-in ports, you will have to buy interface boards.

Almost all laser and dot matrix printers use the parallel ports, and some have both serial and parallel. Many of the daisy wheel printers and most of the plotters use serial ports. For the serial printers, you will need a board with a RS232C connector. The parallel printers use a Centronics type connector. When you buy your printer, buy a cable from the vendor that is configured for your printer and your computer.

Printer sharing

Ordinarily a printer will sit idle most of the time. Some days, I don't even turn my printer on.

Most large offices and businesses own several computers, and almost all of them are connected to a printer in some fashion. In light of the fact that the printer is not used constantly, it would be a waste of money if each computer had a separate printer used only occasionally. Luckily, it's fairly simple to arrange for a printer or plotter to be used by several computers.

If only two or three computers are used and they are fairly close together, this multiple hook-up is not much of a problem. Manual switch boxes costing only from $25 to $150 can allow any one of two or three computers to be switched on line to a printer.

If a simple switch box is used and the computers use the standard parallel ports, the cables from the computers to the printer should be no more than 10' long. Parallel signals will begin to degrade if the cable is much longer than this. A cable system that uses serial signals might be as long as 50'.

If an office or business is fairly complex, then several sophisticated electronic switching devices are available. Some of them can allow a large number of different types of computers to be attached to a single printer or plotter, and many of them have built-in buffers and amplifiers that can allow cable lengths up to 250' or more. The costs here range from $160 up to $1400.

Of course, several networks can connect computers and printers together. Some of the simple networks are rather inexpensive, while some complex ones might be very expensive.

One of the least expensive methods of sharing a printer is for the person to generate the text to be printed out on one computer, record it on a floppy diskette, and then walk over to a computer connected to a printer. If it is in a large office, a single low-cost XT clone could be dedicated to a high-priced printer.

Sources

Just too many laser and plotter companies are around for me to list them all. Look for ads in the local papers and any computer magazine for the nearest dealer.

If Gutenberg were around today, I'm sure that he would be quite pleased with the progress that has been made in the printer business. We've come quite a long way in the last 555 years.

Chapter 12

Telecommunications

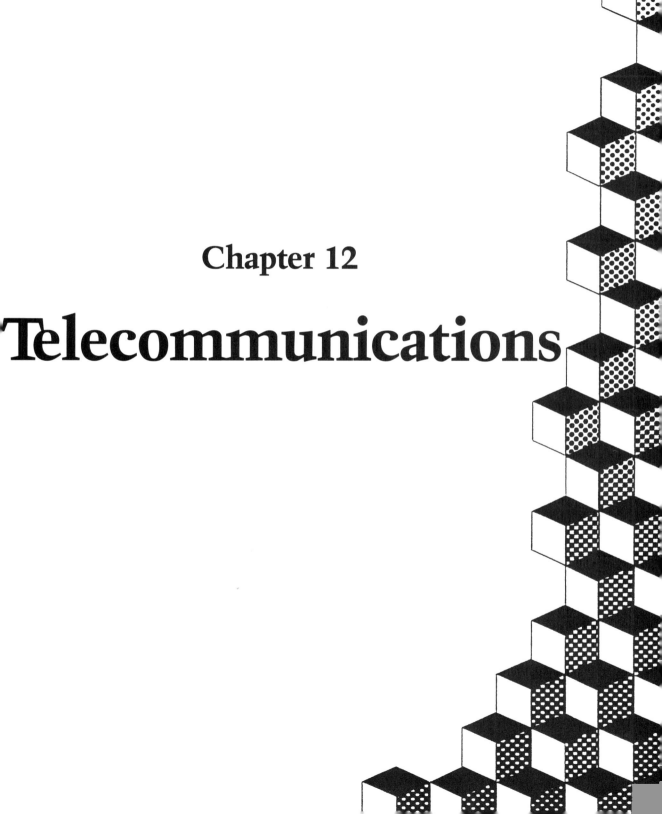

Over 40 million personal computers are installed in homes, offices, and businesses in the US. About 20 million of them have a modem or some sort of communications capability. Modem/communications ability is one of the computer's more important aspects.

A *modem* is an electronic device that allows a computer to use an ordinary telephone line to communicate with other computers also equipped with a modem. Modem is a contraction of the words *modulate* and *demodulate*. The telephone system is analog, while computer data is usually in a digital form. The modem modulates the digital data from a computer and turns it into analog voltages for transmission. At the receiving end, a similar modem will demodulate the analog voltage back into a digital form.

It is easy to use a telephone to communicate with any one of several million persons anywhere in the world. A computer with telecommunications capabilities can just as easily communicate with several million other computers in the world. They can access over 10,000 bulletin boards in the US, take advantage of electronic mail and FAXs, of up-to-the-minute stock market quotations, and of a large number of other on-line services such as home shopping, travel agencies, and many other data services and databases.

Basic types of modems

The two basic types of modems are the *external* desktop and the *internal* desktop. Each type has some advantages and disadvantages.

External

The external modem requires some of your precious desk space and a voltage source. It also requires a COM port to drive it. The good news is that most external models have LEDs that light up and let you know what is happening during your call.

Both the external and the internal models have speakers that let you hear either the phone ringing or (otherwise) the busy signal. Some of the external models have a volume control for the built-in speaker.

Internal

The internal modem is built entirely on a board, usually a half or short one. The good news is that it doesn't use up any of your desk real estate, but the bad news is that it uses one of your precious slots. Also, it does not have the LEDs to let you know the progress of your call. Even if you use an external modem, if your motherboard does not have built-in COM ports, you will need an I/O board and have to use one of your slots for a COM port.

Communications software

In order to use a modem, you must drive and control it with software. Dozens of communication programs can be used. One of the better ones is Relay Silver from

VM Personal Computing, (203) 798-3800. They also publish Relay Gold, one of the most versatile of the high-end communications software packages. It has features that allow remote communications, accessing mainframes, and dozens of utilities not found on the usual communications programs.

Crosstalk, (404) 998-3998, was one of the earlier modem programs. A Crosstalk for Windows now is on the market, working with any Windows version and thus being very easy to learn and use.

ProComm, (314) 474-8461, is one of several low-cost shareware programs. In some ways, it can outperform some of the high-cost commercial programs. The registration cost is $89.

Qmodem, (319) 232-4516, is another excellent shareware program for a registration cost of only $30.

You can get copies of shareware programs from bulletin boards or from any of the several companies who provide public domain software. Shareware is not free. You can try it out and use it, but the developers usually ask that you register the program and send in a nominal sum. For this low cost, they will usually provide a manual and some support (all in all, a very good deal).

Protocols

Protocols are procedures that have been established for exchanging data, along with the instructions that coordinate the process. Most protocols can sense when the data is corrupted or lost due to noise, static, or a bad connection. It will automatically resend the affected data until it is received correctly.

Several protocols exist, and the sending and receiving modems should both use the same protocol. The most popular ones are Kermit (named for Kermit the Frog), Xmodem, and Ymodem. These protocols transmit a block of data along with an error- checking code, wait for the receiver to send back an acknowledgement, and then send another block and wait to see if it got through okay. If a block does not get through, the protocol immediately sends it again. Protocols such as Zmodem and HyperProtocol send a whole file in a continuous stream of data with error checking codes inserted at certain intervals. It then waits for confirmation of a successful transmission. If the transmission is unsuccessful, then the whole file must be sent again.

Baud rate

Telephone systems were originally designed for voice and have a very narrow bandwidth. They are subject to noise, static, and other electrical disturbances. These problems and the state of technology at the time limited the original modems to about 5 characters per second, or a rate of 50 baud.

We get the term *baud* from Emile Baudot (1845-1903), a French inventor. Originally, the baud rate was a measure of the dots and dashes in telegraphy. It is now defined as the actual rate of symbols transmitted per second. For the lower baud rates, it is essentially the same as bits per second. Remember that it takes 8 bits to make a character. Just as we have periods and spaces to separate words, we must use one *start bit* and two *stop bits* to separate the on/off bits into characters. A transmission of 300 baud would mean that 300 on/off bits are sent in one second. Counting the start/stop bits, 11 bits are needed for each character. 300 divided by 11 gives about 27 characters per second (CPS).

Some of the newer technologies might actually transmit symbols that represent more than one bit. For baud rates of 1200 and higher, the CPS and baud rate can be considerably different.

Some fantastic advances have been made in the modem technologies. Most modems sold today operate at 2400 baud, but many are sold that operate at 4800 and 9600 baud. The 9600 baud rate, along with the v.42 four to one compression, will become the standard. Using the 4:1 compression will allow a transmission rate of 38,000 baud.

During modem communication, both the sending and receiving unit must operate at the same baud rate and use the same protocols. Most of the faster modems are downward-compatible and can operate at the slower speeds.

Ordinarily, the higher the baud rate, the less time will be needed to download or transmit a file (although this might not always apply because more transmission errors might occur at higher speeds, demanding that part or all of the file be retransmitted). If the file is being sent over a long distance line the length of telephone connect time can be costly. If the modem is used frequently the telephone bills can be very substantial, especially if you have a slow modem.

How to estimate connect time

You can figure the approximate length of time that it will take to transmit a file. For rough approximations of CPS, you can divide the baud rate by 10. For instance, 1200 baud would be 120 CPS and 2400 baud would be 240 CPS. Look at the directory, determine the number of bytes in the file, and then divide the number of bytes in the file by the CPS. Multiply that figure by 1.3 for the start/stop bits to get a final approximation. For example, to figure the time for a 40K file with a 1200 baud modem, divide 40K by 120 CPS to get 333 seconds and then multiply by 1.3 to get about 433 seconds or 7.2 minutes.

If you transmitted the same 40K file with a 2400 baud modem, it would take 40,000/240 = 167 × 1.3 = 217 seconds or 3.6 minutes to send. With a 9600 baud modem, the same 40K file could be sent in about 55 seconds. If your file was 80K, these figures would double.

What to buy

The modem you buy depends on what you want to do and how much you want to spend. Hundreds of kinds of modems have been manufactured, all with different functions and prices.

One of the most popular early modems was made by Hayes Microcomputer Products. They have become the IBM of the modem world and have established a *de facto* standard. Almost all modems (except the very cheap ones) emulate the Hayes standard.

Many combinations of FAX-modems are available. (See my comments about FAX later in this chapter). If you are involved in any kind of business, you should consider this option.

Considering the telephone rates for long distance, if you expect to do much communicating with a modem, you might want to spend a bit more and buy a high-speed modem. Don't buy anything less than a 2400 baud machine, and get a 9600 baud one if you can afford it. Look in any computer magazine, and you'll see dozens of ads for modems, with most of them being fairly close in quality and function.

One company that I do want to mention is US Robotics. They manufacture a large variety of modems, especially the high-end high speed type. They'll send you a free 110 page booklet that explains about all you need to know about modems. To get this free booklet, just call (800) 342-5877.

Installing a modem

If you are adding a modem on a board to an already assembled system, first remove the computer cover. Then find an empty slot and plug it in.

Set configuration

Before the modem board is plugged into the slot on the motherboard, it must be set or configured to access either serial port COM1 or COM2. The board will have jumpers or small switches that must be set to enable COM1 or COM2. If you have an I/O board in your system with external COM ports or built-in COM ports on your motherboard, you must disable whichever port that will be used for the modem. If the modem board is set to use COM1, then COM1 on the I/O board (or motherboard) must be disabled.

If you are installing an external modem, you must go through the same procedure to make sure the COM port is accessible and does not conflict. If you have a mouse, a serial printer, or some other serial device, you will have to determine which port they are set to. You cannot have two serial devices set to the same COM port.

Plug in the board and hook it up to the telephone line. Unless you expect to do a lot of communicating, you probably will not need a separate dedicated line. The modem might have an automatic answer mode (in which it will always answer the telephone). Unless you have a dedicated line, this mode should be disabled. Check your documentation for the switch or means to disable it.

Plug-in telephone line

Having the modem and telephone on the same line should cause no problems unless someone tries to use the telephone while the modem is using it. If you are going to be using the modem extensively or have a FAX, then you might consider getting a separate dedicated line.

If you have a single line, you might consider buying a routing device. These devices can detect whether the incoming call is voice, for a modem, or for a FAX.

In some areas, you can get a special type of phone service for the telephone rings. When the telephone rings, it can have one ring for modems and FAX and a different and distinctive ring for human voice.

There should be two connectors at the back end of the board. One might be labeled for the line in and the other for the telephone. Unless you have a dedicated telephone line, you should unplug your telephone, plug in the extension to the modem and line, and then plug the telephone into the modem. If your computer is not near your telephone line, you might have to go to a hardware store and buy a long telephone extension line.

A simple modem test

After you have connected all of the lines, turn on your computer and try the modem before you put the cover back on. Make sure you have software. Call a local bulletin board. Even if you can't get through or have a wrong number, you should hear the dial tone and then hear it dial the number.

You might often have trouble determining which COM port is being used by a device. You can use the AT command to determine if your modem is working. At the DOS C prompt, type ECHO AT DT12345>COM1: all in *uppercase*. If the modem is set properly, you will hear a dial tone and then the modem will dial 12345. If two devices are set for COM1, there will be a conflict: the computer will try for a while and then give an error message and the familiar "Abort, Retry, Ignore, Fail?" phrase.

Cables

An external modem will be connected to one of the COM ports with a cable. If you did not get a cable with your unit, you will have to buy one. If you have built-in COM ports, the cable will cost about $5. If you have to use the bus to access the ports, you will need a cable and an I/O board with serial ports.

Bulletin boards

Over 100 bulletin boards (BBS) are active in the San Francisco Bay area, and twice that many are accessible in the Los Angeles area. Some of the bulletin boards are set up by private individuals, while others are set up by companies and vendors as a service to their customers. Some others are set up by users groups and other special interest organizations. In addition, over 100 boards nationwide have been set up for doctors and lawyers alone. The list of bulletin board types is endless: you probably won't be surprised to know that there's a gay bulletin board in the San Francisco area, as well as a few X-rated and dating ones.

One of the largest bulletin boards is Channel 1 in Cambridge, MA. Many bulletin boards are difficult to access because their lines are always busy, so Channel 1 has over 60 telephone lines. They have several gigabytes of public domain software, games, news, weather reports, and other useful services. They offer access from 1200 to 9600 baud. For more information, call (voice) (617) 864-0100.

Most of the bulletin boards are set up to help individuals. They usually have lots of public domain software and a space where you can leave help messages, advertising to sell something, or just plain old chit-chat.

If you are just getting started, you probably need some software. Some public domain software packages are equivalent to almost all of the major commercial programs and, better yet, they're free.

Viruses and Trojan horses

Everyone's heard reports of hidden "viruses" in some public domain and even in some commercial software. This software might appear to work as it should for some time but eventually could contaminate and destroy many of your files. Viruses also often cause the files to grow in size and become larger.

Trojan horses usually do not contaminate other files. Instead, they simply lie dormant for a certain length of time or until a program has been run a certain number of times, and then they destroy the file.

If you download bulletin board software, you should probably run it from a floppy disk until you are sure that it's not "sick." Several companies have developed software to detect viruses, but you can't possibly detect them all.

People who create viruses and other destructive software are sick, mean, and evil-spirited. Apparently, they get their kicks by indiscriminately hurting people who have done them no wrong. I think that it's a crime that deserves severe punishment, such as depriving them of one or more of their most prized appendages. Better yet, we could deprive them of the right to ever own or use a computer (probably more disheartening than the loss of a mere appendage). So far, however, most of these "captured criminals" have been released with little more than probation and a warning.

Fortunately, viruses are not too common, but even a single virus is one virus too many.

Illegal activities

Some bulletin boards have also been used for illegal and criminal activities. For example, stolen credit card numbers and telephone charge numbers have been left on the bulletin boards.

Because of this type of crime, many of the SYSOPS (bulletin board SYStem OPerators) are now carefully checking any software uploaded onto their systems. Many of them also now restrict access to their boards. Some have even had to start charging a fee because of the extra time it takes them to monitor the boards.

Cost to use

At one time, most of the bulletin boards were free of any charge, with the only bill you had to pay being to the phone company (for the calls). Now, however, because of illegal activities and possible virus contaminations, the SYSOPs now have to spend much more time checking the downloads, so they often charge a nominal fee.

One way to reduce phone charges is to use data compression. Bulletin Boards have been using a form of data compression for years. Of course, data compression allows them to store more data on their hard disks. Several public domain programs squeeze and un-squeeze the data.

Where to find the bulletin boards

Several local computer magazines devote a lot of space to bulletin boards and user groups. In California, the MicroTimes and Computer Currents magazines have several pages of bulletin boards and user groups each month. The Computer Shopper has the most comprehensive listing of bulletin boards and user groups of any national magazine.

On-line services

Several large national bulletin boards have been created as information and reference services (such as CompuServe, Dataquest, Dow Jones, and Dialog). These companies have huge information databases that a caller can search and download information from as easily as pulling the data off his/her own hard disk. These companies charge a fee for the connect time.

Prodigy is unlike the other on-line services because it doesn't charge for connect time (just a very nominal monthly rate). They have phone service to most areas in the larger cities so that there is not even a toll charge. Their impressive list of services includes home shopping, home banking, airline schedules and reservations, stock market quotations, and others. Prodigy is a real bargain. Although it's relatively slow, the price more than compensates for it. You can contact Prodigy at (800) 759-8000.

E-mail

Many of the national bulletin boards offer electronic mail or E-mail along with their other services. These services can be of great value to some individuals and businesses.

E-mail subscribers are usually given a "post office box" at these companies, which is usually a file on one of their large hard disk systems. When a message is received, it is recorded in this file. The next time the subscriber logs on to this service, s/he will be alerted that there is "mail" in their in-box.

E-mail is becoming more popular every day, and now there are several hundred thousand subscribers. The cost for an average message is about one dollar (compared to the cost for overnight mail with the US Post Office, Federal Express, or UPS, which could range from $11 – 13).

Here are some companies that provide E-mail at the present time:

AT&T Mail	(800) 367-7225
CompuServe	(800) 848-8990
DASnet	(408) 559-7434
MCI Mail	(800) 444-6245
Western Union	(800) 527-5184

Banking by modem

Many banks offer systems that let you do all your banking with your computer and a modem from the comforts of your home. You would never again have to drive downtown, hunt for a parking space, and then stand in line for half an hour to do your banking.

Public domain and shareware

If you don't have a modem yet or if the local bulletin boards don't have the software you need, several companies will ship you public domain and shareware software on a floppy diskette. These companies have thousands of programs and usually charge from $3 – 24 for a disk full of programs. Most of the public domain software companies advertise in the computer magazines listed in Chapter 16. They will send you a catalog listing of their software, for which some might charge you a small fee. Call them for details and latest prices.

ISDN

ISDN is an acronym for Integrated Services Digital Network. Eventually the whole world will have telephone systems that use this network, which will be able to transmit voice, data, video, and graphics in digital form rather than the present analog. When this happens, we can scrap our modems because we'll only need a simple interface to communicate.

ISDN is already installed in several cities and is scheduled to be fully implemented in the US by 1992. Don't throw away your modem yet, though. The new service will be rather expensive and might not be available at all locations for some time.

Facsimile boards and machines

Facsimile (FAX) machines have been around for quite a while; newspapers and businesses have used them for years. They were the forerunners of the scanning machines.

The early machines were similar to the early acoustic modems. Both used foam rubber cups that fit over the telephone receiver-mouthpiece for coupling. They were very slow and subject to noise and interference. Fax machines and modems have come a long way since those early days.

A page of text or a photo is fed into the facsimile machine and scanned. As the scanning beam moves across the page, white and dark areas are digitized as 1s and 0s and then transmitted out over the telephone lines.

Modems and facsimile machines are quite similar and related in many respects. A modem sends and receives bits of data, while a FAX machine or board usually sends and receives scanned whole page letters, images, signatures, etc. A computer program can be sent over a modem but not over a FAX. A FAX sends and receives the information as digitized data, while a modem converts the digital information that represents characters into analog voltages, sends it over the line, and then converts it back to digital information.

Sometimes one or the other is needed. Both units would not be in use at the same time, so the same phone line can be used for both of them.

Millions of facsimile machines are being used today. Almost all businesses can benefit from using a FAX, which can be used to send documents that include handwriting, signatures, seals, letterheads, graphs, blueprints, photos, and other types of data around the world, across the country, or across the room to another FAX machine.

It costs from $8.50 – 13.00 to send an overnight letter; E-mail can send the same letter for $1. A FAX machine can deliver it for about 40 cents and do it in less than three minutes. Depending on the type of business and the amount of critical mail that must be sent out, a FAX system can pay for itself in a very short time.

Standalone FAX Units

Several facsimile machines are still stand-alone devices that attach to a telephone. They have been vastly improved in the last few years, so most of them are as easy to use as a copy machine. In fact, most of them *can* be used as a copy machine.

Some overseas companies are making fairly inexpensive stand-alone units for as little as $400. You might not be happy with the low-cost ones. You will be better off if you can spend a bit more and get one that uses plain paper and has a paper cutter, a high resolution, a voice/data switch on the system, a document feeder, an automatic dialer, an automatic retry, a delayed transmission, a transmission confirmation, polling, a built-in laser printer, and a large memory. You might not need or be able to afford all of these features, but try to get a machine with as many as possible. Of course, the more features, the higher the cost.

FAX computer boards

Several companies have developed FAX machines on circuit boards that can be plugged into computers. Many of the newer models have a modem on the same board. Follow the same procedure to install a FAX board as outlined earlier for an internal modem.

Special software allows the computer to control the FAX boards. Using the computer's word processor, letters and memos can be written and sent out over the phone lines. Several letters or other information can be stored or retrieved from the computer hard disk and transmitted. The computer can even be programmed to send the letters out at night when the rates are lower.

The computer FAX boards have one disadvantage: they cannot scan information unless there is a scanner attached to computer. Without a scanner, the information that can be sent is usually limited to that which can be entered from a keyboard or disk. As I pointed out before, stand-alone units scan pages of information, including such things as handwriting, signatures, images, blueprints, and photos. It is fairly easy to attach a scanner to your computer. (See Chapter 10 for detailed information on scanners.)

The computer can receive and store any FAX sent. The digitized data and images can be stored on a hard disk and then printed out on a printer.

The FAX boards might cost from $100 up to $1000, all depending on the extras and goodies installed. Pay close attention to the ads and specifications. I have seen several FAX boards advertised for less than $100, but the reason behind the low cost is because these boards will only send FAX files, not receive them. Of course, nowhere in the ad did it mention this fact. Even when I called one company, they were reluctant to admit that their boards only worked in the send mode. The reason they could advertise and sell them for less is because more electronics are required to receive a FAX than to send one. For a boiler room type operation that does nothing but send out ads, this board would probably be sufficient.

Single board modem and FAX

I have an older Intel Connection CoProcessor FAX board. The current model is called SatisFAXtion and is a sophisticated FAX board with its own 10MHz 80186 processor and 128K of RAM. Its 2400 baud modem can use the MNP 5 protocol to effectively send at 4800 bps. Because it has its own processor, it can send and receive FAX messages or operate the modem in the background while the computer is busy on other projects.

My older Connection CoProcessor cost over $1000. The SatisFAXtion, with modem, lists for $499. The street price will be considerably less than this. For more information, contact Intel at (800) 538-3373.

Several other companies manufacture a single board FAX and modem. Look through any of the computer magazines for ads. Before ordering, be sure to ask for complete specifications and check them closely. Some of the less expensive boards might not do everything you want to do.

The Image Communications Company, (800) 666-2496, has a small half card with a Fax and a modem. It makes extensive use of VLSI technology. My Intel Fax and modem Connection CoProcessor is a full length board full of electronics. The Image Communications Company have been able to cram about as much utility onto this small board as I have on my large board. The Image Communications board sells for just $139.

The Complete PC company, (408) 434-0145, has several types of special boards. One has a FAX, a modem, and a telephone answering machine and voice mail. The suggested retail list price is $699.

Installing a FAX board

Most fax boards are very easy to install and easy to operate. If your computer is already assembled, just remove the five screws that hold on the cover. Check your documentation and set any necessary switches. These switches and jumpers will allow you to choose an I/O interrupt that doesn't conflict with any other item on the bus. Then plug the board into an empty slot, replace the cover, and connect the telephone line.

You should have received some software necessary to control the FAX. This software should be installed on your hard disk. You should then be up and ready to send and receive FAXs.

If you use a word-processing program (such as WordStar), in order to create letters or text for a FAX transmission, the text must be changed into an ASCII file before transmission. Most word processors have this capability.

Telecommuting

One reason I took early retirement from my job at Lockheed was because I hated being stuck in commute traffic. Of course, another reason was that my books were selling well.

Millions of people risk their lives and fight the traffic every day. Many of these people have jobs that could allow them to stay home, work on a computer, and then send the data to the office over a modem or a FAX. Even if the person had to buy their own computer, modem, and FAX, it still might be worth it. You could save the cost of gasoline, auto maintenance, and lower insurance, as well as your life (thousands are killed on the highways) telecommuting can be a life-saver.

Being able to work at home would be ideal for those who have young children, for the handicapped, or for anyone who hates traffic.

A good FAX and modem system will pay for itself in a very short time in any business that does a lot of communicating.

More help

The world of communications is vast and technology is advancing at a fantastic pace. I have only touched the surface in this short chapter. Many good books, though, have been published on the subject. One of the more comprehensive ones is *Dvorak's Guide to PC Telecommunications*, published by Osborne McGraw-Hill. It contains just about everything that one could ever want to know about telecommunications.

Chapter 13

Windows

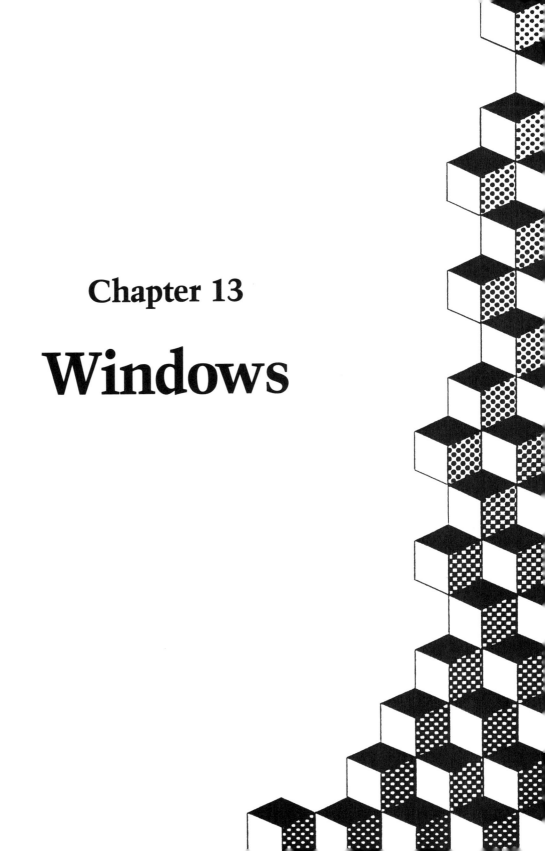

M‎acintosh computers are fairly easy to use; just grab the mouse, point it somewhere on the screen, and click the buttons. The DOS world has been considerably more difficult to live in, but now that we have Windows 3.0 to play with, life will get easier.

DOS software has never been as easy to learn or use as the advertisers and public relations people claim. Actually, you can more easily assemble a computer than learn some of the DOS software packages. Many of the programs are difficult and take time to learn. Some people might read the PR hoopla about how easy a program should be to learn and then feel stupid because they're having trouble.

A need for standards

Computers should be similar to the auto industry. If you know how to drive an automobile, you can drive almost any car without any special training. This concept should be the same for computer software programs, especially for such basic programs as word processing, databases, and spreadsheets.

Of course, some applications must be trained for. Just because a person knows how to drive an automobile doesn't necessarily mean that s/he could handle an 18-wheel semi without a bit of training. Likewise, just because a person knows a word processing program wouldn't necessarily mean that s/he should be able to run complex programs such as Ventura or AutoCAD without a bit of training. Still, it sure would help if more of the software operation was standardized.

Many of the high-cost user friendly programs are supposed to be very easy to learn and use, but many of them aren't. Numerous companies make a lot of money by holding training seminars on these programs, which is ridiculous. In my opinion, if a person pays $600 – $700 for a piece of software, s/he should not have to pay another $600 – $700 to learn how to use it.

Windows 3.0

Microsoft introduced Windows 3.0 on May 22, 1990. Within the first year, it sold about five million copies (which makes it one of the fastest selling software packages ever).

Windows 3.0 is a giant step toward standardization. Unfortunately, most of our present software cannot take advantage of its many features. However, software developers are working overtime to revise and upgrade thousands of different applications that will conform to the Windows 3.0 requirements. Hopefully, this is the beginning of a standard for the DOS PC world.

GUI

Windows 3.0 is a graphical user interface (GUI, pronounced "gooey"), which means that it's a shell that allows other programs and applications to run from within it. You can also run several programs at the same time. For instance, you might be writing a report with a word processor and need data from a spreadsheet. Ordinarily, you would have to close your word processor and open the spreadsheet files. With Windows, you could leave the word processor running, open the spreadsheet, find the data you need, and just import it to the word processor. This excellent tool can increase your productivity and generally make your computing life a lot easier.

One reason the Apple Macintosh is so easy to use is that most of the software it uses operates with icons and a mouse. You need only point at an icon and click instead of learning a bunch of arcane commands. Apple developed the icon system from research done by Xerox at their Palo Alto Research Center (PARC) in the late 1970s. It was a technology that was years ahead of its time, so Xerox did not pursue it (unfortunately for them).

It's been years in the making, but now Windows 3.0 has caught up with the Macintosh and surpassed it.

Why Windows is easier to learn and use

When designing Windows, Microsoft hired several cognitive psychologists to study how people learn and remember. One of the main problems with using a computer is learning and remembering the commands. A person can more easily recognize an object than recall a name. Face it: you can probably recognize the faces of hundreds of people but not be able to remember their names.

A person finds it much easier and faster to point a mouse at an object and click it than to remember an arcane textual command. If you don't use a textual command several times every day, you can have even more difficulty remembering it; conversely, you can easily recognize an icon.

Requirements

You can run Windows 3.0 on an XT in real mode but it will be very slow. Windows runs fairly well on a 286 with a minimum of 1M but will run best on a 386 or 486 with 2 – 4M of RAM.

Windows will run on monochrome monitors, but you would be much happier with color. Windows comes with an excellent Solitaire card game, which I tried to play on a monochrome system—it was almost impossible to do. I don't normally play computer games, but this one is great—it alone can justify the cost of a color monitor.

Automatic setup and installation

Windows 3.0 has many features that make it easy to learn and to use. Even the installation is easy and automatic. The Setup program detects the users hardware configuration and installs Windows to take full advantage of it.

On-line help

Windows has over 1M of hypertext help available. Help is available from within the many applications and the user shell. Pressing the F1 key brings up a screen full of help, which is indexed so that you can quickly find what you need. Help also features simple error messages if you make a mistake, sometimes even offering solutions to the error.

Breaking the 640K barrier

Many applications written for Windows will let you break the 640K barrier and take advantage of the protected or the 386 enhanced mode, which enables the application to use all of the extended memory that you have.

It also speeds up programs, improves handling of high resolution bit-mapped graphics, and allows multitasking by letting you run several application programs at the same time.

Operational modes

Windows 3.0 can run in three different modes: real, standard or protected, and 386 enhanced. See pages 120–122 in Chapter 9 for a description of the these three modes.

Other features of windows

Windows 3.0 comes with many accessories and applications.

Dynamic Data Exchange (DDE)

DDE provides the user a direct data tie with other Windows applications that support DDE. With DDE, if a spreadsheet file is changed and a second file has similar data, the second file is updated automatically. DDE can update all files, even if they are running in several different windows.

Calculator

This calculator is both standard and scientific. Calculations can be stored in memory.

Calendar

The calendar is a combined month-long one, as well as a daily appointment book. Appointments can be set up so that an alarm can flash or a beep can alert you.

Cardfile

The cardfile resembles a set of index cards that can be used to keep track of names, addresses, phone numbers, directions, or other things that you might want. It also can sort itself.

Clock

The clock can be placed in a window and can be displayed as a digital or an old-fashioned analog with hands.

Notepad

The notepad is a text editor for jotting down notes and short memos. It can also be used to create batch files.

Recorder

The recorder allows you to record a sequence of keystrokes and mouse actions to create macros. You can record up to several hundred keystrokes or mouse clicks, and then have them played back or inserted in a file by using only two or three more keystrokes.

Clipboard

The clipboard is a temporary storage location always available for transferring information or data from one window or application to another. Information is cut or copied from one application and then pasted into another location or window. If you wanted to move data from a spreadsheet or a database to a word processor, the data could be imported to the clipboard and then moved from the clipboard to the word processor.

Write

Write is a basic word processor that can be used to create, save, and print documents. It is not as full-featured as a stand-alone word processor but can be quite handy for many small tasks.

Paintbrush

With Paintbrush, you can create simple drawings or elaborate color art and graphics. It has several icons that represent tools that can help you create your drawings.

Terminal

The terminal is an application that allows you to connect to another computer over a serial cable. It also can be the operating software for a modem.

Reversi and Solitaire

Reversi and Solitaire are games. If you have a "good" boss, you might be able to convince him/her that playing these games is necessary to learn the basics of Windows. If s/he is not all that understanding, you'll have to resort to playing them when s/he is not around. If you should see him/her coming, the games can be quickly removed from the screen and replaced with a legitimate job that you had loaded and running in the background.

I should warn you that Solitaire (in both graphics and playability) is addictive. I don't usually play games on a computer, but I had to try out the Solitaire game in order to write about it. I won the first time I played it and then couldn't stop playing for another hour.

Microsoft has several other games bundled separately if you want to buy them.

Wallpaper

If you are tired of looking at a plain old screen when you turn on your computer, you can use Windows to cover it with "wallpaper." Windows comes with several bitmapped files with the extension .BMP. To use them, load Windows. After that, from inside the Program Manager, use your mouse to double click on the Control Panel icon. Then double click on the Desktop icon, choose Wallpaper, and choose the file you want to use.

Windows vs. OS/2

Windows 3.0 has almost all of the features found in OS/2. Windows 3.0 costs about $75 at discount houses, while OS/2 costs from $300 to $700. Windows 3.0 will operate on all clones, while OS/2 is designed for high-end workstations and might have to be customized for the system. Windows 3.0 already has an installation base of several million but at this time, only about 200,000 OS/2 systems have been sold. In light of all this, which area do you think the software developers will be writing for? Without a doubt, Windows 3.0 will be the GUI standard for years to come.

This is very good news for all those who have had trouble learning to use a computer. In the past, people bought Macintosh because it was easier to use, even though it cost about twice as much as a clone. Now you don't have a reason to buy a Macintosh, unless you happen to be an absolute Macintosh fanatic.

Microsoft Windows
16011 N.E. 36th Way
Redmond WA 98073
(206) 882-8080.

The list price is $149, but the street price is about half that.

Applications for windows

Hundreds of third party application programs have been developed to run in the Windows environment, and more are being developed daily. They take advantage of the windows and pull down menus of this vital tool. I'll tell you about a few of them.

YourWay

YourWay acts as a core and allows users to build their own personal productivity workstation. When Windows is called up, YourWay comes up and is used as a control center for other programs. It allows you to launch into other programs such as spreadsheets, databases, or word processors. It allows data exchange through the Windows Clipboard and by traditional file import and provides the user with a direct data tie with other Windows applications that support Dynamic Data Exchange (DDE). With DDE, if a spreadsheet file is changed and a second file has similar data, the second file is updated automatically.

For more information, call Prisma Software at (619) 259-1400.

Norton Desktop for Windows

This software package contains several utilities to work under Windows, including all the old familiar Norton Utilities, plus Backup, batch file Builder, and a comprehensive manager for windows files and directories. It's available from Symantec, (408) 253-9600.

Micrografx

Windows Graph is a presentation graphics program that uses Windows. Data from spreadsheets and other files can be ported to the program to make all sorts of graphs. Micrografx also has several other graphics programs and graphics aids such as ClipArt, Draw Plus, Designer, and Graph Plus. Charisma is a large graphics program that integrates most of these programs into a single package, making an excellent tool for creating all kinds of graphics needed for presentations, designs, and desktop publishing.

    Call Micrografx at (800) 272-3729.

Ventura Publisher

Ventura Publisher is the premier program for desktop publishing and is now much easier to use under Windows.

    Call Ventura Software at (800) 822-8221.

PageMaker

PageMaker rivals Ventura as a powerful desktop publishing program. It was originally designed for the Macintosh but now works in the DOS world.

    Call Aldus Corporation at (206) 622-5500.

Crosstalk for Windows

You'll never have had an easier time communicating with your modem. You can use your mouse to open your phone book and place calls.

    Call DCA, Inc., at (800) 241-4762.

Windows Express

The hDC Computer Corporation has developed several applications for Windows. Windows Express allows you to customize Windows and your applications, design your own icons and help screens, password protect confidential files, as well as perform other useful things.

    Call hDC Computer Corporation at (206) 885-5550.

Dragnet and Prompt

Dragnet is a text-retrieval software, able to search your entire disk for a file, a phrase, or a word.

    Prompt can search your disk for files and then let you view, copy, rename, delete, compress, or encrypt them. Prompt can operate alone or be linked with Dragnet.

    Call Access Softek at (415) 654-0116.

Word processors

Several word processors now operate under Windows, one of them of course being Word for Windows from Microsoft. WordPerfect, WordStar, AMI, and several others also have Windows-functional versions out. I've described some in more detail in Chapter 14.

GeoWorks

Although you can run Windows 3.0 on an XT, I don't recommend it. The XT just doesn't have enough power; it's very slow and offers very little advantages over plain old DOS. However, GeoWorks has developed a "poor man's" GUI that works great on the lowly XT, bringing over all of the Windows advantages and allowing true multitasking, scalable fonts, a mouse-driven graphical user interface, and an exceptional printer output.

Of course, GeoWorks works even better with the larger machines than it does with the XT.

At the present time, only a limited number of software applications are available for GeoWorks. Most of them were developed by GeoWorks and are bundled with the program, among them being GeoManager, GeoDraw, GeoWrite, GeoComm, GeoPlanner, GeoDex, a notebook, a calculator, and a scrapbook. Several third-party developers are creating new applications, so many more should be available by the time you read this.

The GeoWorks Ensemble, which includes all of the applications listed above, lists for $195—just a bit more than Windows 3.0.

Order from

GeoWorks
2150 Shattuck Ave.
Berkeley, CA 94704
(415) 644-0883

New windows programs

I listed only a few of the hundreds of programs that are available for Windows because I couldn't possibly list them all. Windows programs have been written for almost every application that you can dream about, and more being developed every day. Look through the computer magazines for ads, articles, and reviews.

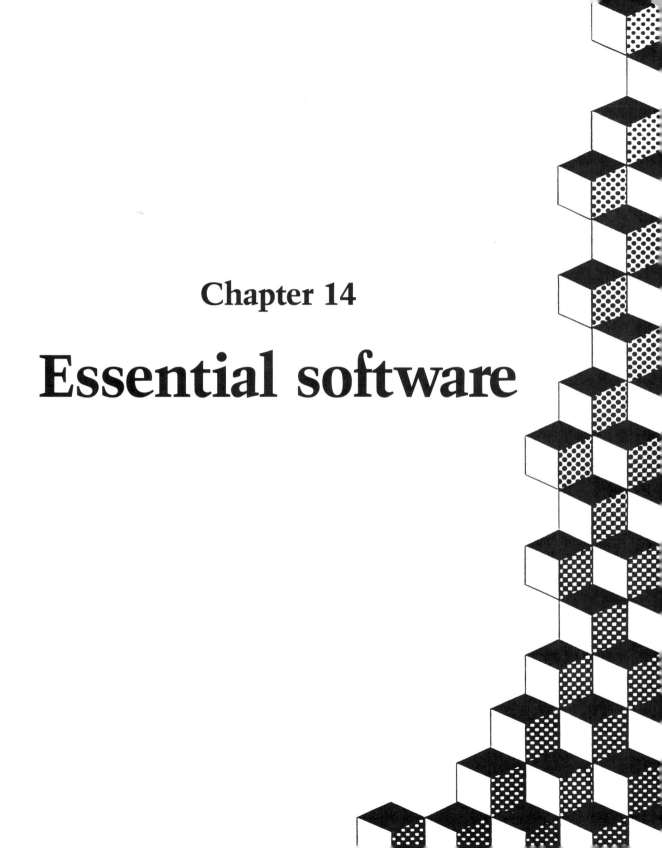

Chapter 14
Essential software

All of the software I'll describe here can be used in business or at home. More software exists, already written and immediately available, than you can use in a lifetime. Except for very unusual applications, the ordinary user should never have to do any programming. Off-the-shelf programs can do almost everything that you could ever want to do with a computer. Yet thousands and thousands of software developers are working overtime to design new programs. Almost like the soap business, they constantly issue new and improved versions when, quite often, the old version can still do all you need to do. I know many people who are still using dBASE II, which is over ten years old.

For most general applications, you'll only need certain basic programs. Speaking of basic, BASIC is one that you'll need. GW-BASIC from Microsoft is more or less the standard and is included with all versions of MS-DOS. Many applications still use BASIC. Even if you are not a programmer, the language is simple enough that anyone can design a few special applications with it.

The several categories of programs that you will need include a Disk Operating System (DOS), Word Processors, Databases, Spreadsheets, Utilities, Shells, Communications, Windows, Graphics, and Computer Aided Design (CAD). Depending on how you intend to use your computer, hundreds of other programs can fulfill special needs.

Software can be more expensive than the hardware, and the prices can also vary from vendor to vendor. Few people pay the list price; it will pay you to shop around. I have seen software with a list price of $700 that was advertised by a discount house for $350. You should also remember that some excellent free public domain programs can do almost everything that the high-cost commercial programs can do. Check your local bulletin board, User Group, or the ads for Public domain software in most computer magazines. Some excellent Shareware programs also can be registered for a very small sum.

I can't possibly list all of the thousands of software packages available. Again, subscribe to the magazines listed in this book; most of them have detailed reviews of software in every issue.

Operating systems software
MS-DOS 5.0

DOS to a computer is like gasoline to an automobile; without it, the system won't operate. DOS is an acronym for Disk Operating System, but it does much more than just operate the disks. In recognition of this fact, the new OS/2 has dropped the D.

You can use any version of DOS on your computer, even MS-DOS 1.0 (although I don't know why anyone would want to do that, considering how much it would limit you). Personally, I recommend that you buy MS-DOS 5.0 or DR DOS 6.0.

When the operating system, CONFIG.SYS, buffers, drivers, TSRs, and other things are loaded into the 640K of available RAM, often not enough memory is left to load other programs. DR DOS 6.0 and MS-DOS 5.0 both take the operating system, TSRs, buffers, drivers, and other programs out of low memory and place them in memory above 640K. The older versions of DOS just load everything into low memory.

DOS contains many good commands; but, if not used properly, some of them can be disastrous. Be very careful when using commands such as FORMAT, DEL, ERASE, COPY, and RECOVER. When invoked, RECOVER renames and turns all of the files into FILE0001.REC, FILE0002.REC, etc. The disk will no longer be bootable, and critical files might be garbled. Many experts say that you should erase the RECOVER command from your disk files and leave it only on your original diskettes, to be used only as a last resort. The Norton Utilities, Mace, PC Tools, or one of the other utilities would be much better to UnErase or restore a damaged file.

DOS has about 50 different commands and files. You might never have a need or use for the majority of them. The ones I use most are COPY, DEL, .BAT, REN, FORMAT, DIR, MD (for Make Directory), RD (for Remove Directory), CD (for Change Directory), TYPE, EDITOR, and CHKDSK. These few commands will allow one to do most of the ordinary computing tasks.

DR DOS 6.0

The DR stands for Digital Research, not for Doctor. The Digital Research Corporation was founded by Gary Kildahl, the developer of CP/M—the first operating system for personal computers.

DR DOS 6.0 is completely compatible with MS-DOS. You could say that it is a clone of MS-DOS, except for the fact that DR DOS 6.0 preceded MS-DOS 5.0 by a couple of years.

DR DOS 6.0 has several good features. For instance, FileLINK allows you to connect and transfer files over a serial cable. ViewMAX allows you to view, organize, and execute files and commands using only one or two keystrokes or a mouse. DR DOS has comprehensive on-line help, supports hard disk partitions up to 512M, has Disk cache, a full-screen text editor, password protection, and many other features not found in MS-DOS.

If you work in a large office, you might want to keep some of your nosy neighbors from snooping into your personal files; DR DOS 6.0 allows you to protect them with a password. The ViewMAX shell feature (mentioned before) lets DR DOS use icons, similar to the operation of Windows 3.0. Thus, you can use a mouse to quickly open files, copy, delete, and perform many of the other commands and functions. The feature also has a clock that can run in a window of the screen. Also featured is a calculator for on-screen calculations.

DR DOS is easily installed; just copy it onto your hard disk. You don't even have to reformat your disk; if you have a previous version of DOS on the disk, DR DOS will copy over and replace it. DR DOS has the same commands as DOS and is fully compatible with it.

Call Digital Research, (800) 443-4200, for more information.

OS/2

OS/2 is designed specifically for high-end applications and large workstations. Its ROM BIOS functions might not be compatible with most clones. Companies such as Compaq, Dell, Zenith, and others have designed unique OS/2 versions for their clones. Unless you have some special need for multitasking or multiusing, you might not need OS/2; you can probably get by with MS-DOS and Windows 3.0. Besides, not much software is available yet that takes advantage of OS/2. Also, it might cost two or three times more than DOS.

DESQview

DESQView resembles Windows in some respects in that it runs on top of DOS and allows multitasking and multiusers. You can have up to 50 programs running at the same time and have as many as 250 windows open. DESQview runs all DOS software and is simple to learn and use. Early versions of DESQview caused a conflict in memory when run under Windows 3.0, but new versions now run under Windows 3.0 with no problems.

Call Quarterdeck Office Systems, (213) 392-9851, for information.

Concurrent DOS 386

Concurrent DOS 386 is also from Digital Research and is another excellent alternative to OS/2. It's a multitasking and multiuser operating system and takes advantage of the 80486 virtual 8086 mode, allowing simultaneous processing of DOS applications. It's easy to install and operates with familiar DOS commands.

Call Digital Research, (800) 443-4200, for information.

DOS help programs

Learning DOS can be very difficult, so several Help programs have been developed. One such program is Doctor DOS from VMG, (405) 722-8244. The Phoenix Company, (617) 551-4000, one of the first developers of a clone BIOS, also has an excellent Help program for learning DOS. Many others are available, even among public domain programs. Look for ads in computer magazines.

Word processors

The largest focus of all software is on word processing. Literally, hundreds of word processor packages have been developed, each one slightly different than the others. It amazes me that companies can find so many different ways to do the same thing. Most of the word processor programs come with a spelling checker, and some come with a thesaurus (which can be very handy). They usually also include several other utilities for such things as communications programs for your modem, outlines, desktop publishing, print merging, and many others.

As always, many good word processors other than those listed are available. Look for ads and reviews in computer magazines.

WordStar

I started off with WordStar on my little CP/M Morrow with a hefty 64K of memory and two 140K single-sided disk drives. It took me some time to learn it. I have tried several other word processors since then and found that most of them would require almost as much time to learn as WordStar did originally. I don't have a lot of free time; but because WordStar does all I need, I haven't learned too many other processors.

Probably, more copies of WordStar exist than any other word processor. Most magazine and book editors expect their writers to send manuscripts to them on a diskette in WordStar. Newer versions work under Windows. Call WordStar, (800) 227-5609, for details.

WordStar also has an educational division that offers an excellent discount to schools, both for site licenses and for student purchases. You can reach the educational division at (800) 543-8188.

WordPerfect

WordPerfect is one of the hottest-selling word processors, so it must be doing something right. One thing I know they do right is give free unlimited support. WordPerfect has the ability to select fonts by a proper name, has simplified printer installation, the ability to do most desktop publishing functions, columns, plus other useful functions and utilities. WordPerfect now works with Windows.

Call WordPerfect, (801) 225-5000, for information.

Microsoft Word for Windows

Word for Windows was developed by the same people who gave us MS-DOS and lets you take advantage of all features and utilities of Windows. It is among the best sellers in the country. If you have previously learned a different word processor, Word for Windows includes a manual listing the differences in most popular word processors, helping you quickly become accustomed to Word for Windows.

Call Microsoft Corp., (206) 882-8080, for more information.

AMI

AMI is a very good word processor that will do most of the things needed for desktop publishing. Plus, it works with Windows.

Call Samna at (800) 831-9679 for more information.

PC-Write

PC-Write is the least expensive of all the word processors, considering that it's shareware and thus free if copied from an existing user. Quicksoft asks for a $16 donation, and sells full registration (including a manual and technical support) for $89. PC-Write is easy to learn as well as an excellent personal word processor.

For more information, call Quicksoft, Inc., at (800) 888-8088.

Grammar checkers

You might be the most intelligent person alive, but you might not be able to write a simple intelligible sentence. Several grammar checking programs are available that can work with most of the word processors. They can analyze your writing and suggest ways to improve it. Here are the phone numbers for two of them:

Right Writer Que Corporation (800) 992-0244
Grammatik   Reference Software (800) 872-9933

Database programs

Database packages are very useful for business purposes, allowing you to manage large amounts of information. Most programs allow one to store information, search it, sort it, do calculations, and make up reports as well as perform other functions. At the present time, almost as many database programs are out as there are word processors. Unfortunately, few of them are compatible with others. However, the industry is laboring to establish some standards under the Structured Query Language (SQL) standard, and several of the larger companies have announced their support for this standard. The average price for the better known database packages is almost twice that of word processors.

dBASE IV

Ashton-Tate, with dBASE II, was one of the first companies to develop a database program for the personal computer. dBASE is a very powerful program and has hundreds of features, but it's also highly structured and can be a bit difficult to learn. dBASE IV is much faster than dBASE III, has a built-in compiler, SQL, and an upgraded user interface along with several other enhancements.

For more information, call Ashton-Tate at (213) 329-8000.

askSam

The funny looking name is an acronym for Access Knowledge via Stored Access Method. It is a free-form, text-oriented database management system and is almost like a word processor. Data can be typed in randomly and then sorted and accessed. Data can also be entered in a structured format for greater organization. askSam is not quite as powerful as dBASE IV but is much easier to use. It is also much less expensive, being ideal for personal use and for the majority of business needs.

Seaside also has discount programs for students. Students can buy the $295 program for only $45 when the order is placed by an instructor. Better yet, any instructor who places an order for ten or more copies will get a free copy (a fantastic bargain).

For more information, call Seaside Software at (800) 327-5726.

R:BASE 3.1

R:BASE has been around for a long time but has now been revised and updated. It has pull-down menus, mouse support, is fully relational for multi-table tasks, and has an English-like procedural language. R:BASE is one of the more powerful and more versatile of the present-day database programs. Microrim is so sure that you will like the program that they offer an unlimited, no- questions-asked, money-back, 90-day guarantee.

For more information, call

Microrim
3925 159th Ave. N.E.
Redmond, WA 98052
(206) 885-2000.

FoxPro

FoxPro is very easy to use, having windows and able to be controlled by a mouse or the keyboard. Of course, using FoxPro with a mouse saves several keystrokes. FoxPro also has several different windows and a View Window in the master control panel to create databases, open files, browse, set options and other functions. You don't have to be a programmer to type commands into the Command Window to operate FoxPro; the Browse Window lets you view, edit, append, or delete files. FoxPro also has Memo Fields, a built-in editor, and allows you to create Macros as well as use extensive context-sensitive help, plus more.

For more information, call Fox Software, (419) 874-0162.

Paradox

Paradox is fairly easy to learn and use, as well as being fast and powerful. It is designed for both beginners and expert users and is a full-featured relational database that can be used on a single PC or on a network. The main menu has functions like View, Ask, Report, Create, Modify, Image, Forms, Tools, Scripts, and Help. Choosing one of these items will bring up options associated with that item. The function keys are used extensively.

The query by example is very helpful for beginners and experts alike. Paradox has a very powerful programming language—PAL. Experienced programmers can easily design special applications.

Paradox is one of the Borland family of products. Philippe Kahn is the founder and president of Borland. He is young—about the same age as Gary Kildahl, Bill Gates, Steve Jobs, Wozniak, and several of the other young computer pioneers. He was penniless when he came to this country from Belgium, but he got a little wealthier when he first developed Sidekick, then Turbo Pascal, and then pretty soon dozens of products. His company has recently acquired Ashton-Tate and dBase IV.

For more information, call Borland International at (408) 438-5300.

Spreadsheets

Spreadsheets are primarily number crunchers. They have a matrix of cells in which data can be entered, and data in a particular cell can be acted on by formulas and mathematical equations. If the data in the cell acted on affects other cells, recalculations are done on them also.

Several of the tax software programs use a simple form of spreadsheet. The income and all the deductions can be entered. Then, if an additional deduction is discovered, it can be entered and all the calculations will be done over automatically.

In business, spreadsheets are essential for inventory, expenses, accounting purposes, forecasting, making charts, and dozens of other vital business needs. Because such a large number of spreadsheet programs are available, I'll only mention a few of them. For other spreadsheet programs, check the ads and reviews in computer magazines.

Microsoft Excel

For years, Lotus 1-2-3 has been the premier spreadsheet, but it appears that Excel will take the top spot and honors. Excel is a very powerful program, with pull-down menus, windows, and dozens of features. It can even perform as a database.

For more information, call Microsoft at (206) 882-8080.

Quattro

The Quattro spreadsheet looks very much like Lotus 1-2-3 but has better graphics capabilities for charts, calculates faster, has pull-down menus, can print sideways, plus several other things Lotus 1-2-3 can't do. Of course, one of the better features is the suggested list price of $195 (only $148 from a discount house).

For more information, call Borland International at (408) 438-8400.

SuperCalc5

SuperCalc, introduced in 1981, was one of the pioneer spreadsheets. It has never enjoyed the popularity of Lotus, though it has features not found in Lotus. It is compatible with Lotus 1-2-3 files and can link to dBASE and several other files and is also an excellent spreadsheet.

Computer Associates has also developed several excellent Account packages costing from $595 to $695.

For more information, call Computer Associates at (408) 432-1727.

Utilities

Utilities are essential tools that can unerase a file, detect bad sectors on a hard disk, diagnose, unfragment, sort, and do many other things. Norton Utilities was the first (and still foremost) in the Utility department. Mace Utilities has several functions not found in Norton. Mace Gold is an integrated package of utilities that includes POP, a power-out protection program, a Backup up utility, TextFix, and dbFix for data retrieval. PC Tools has even more utilities than Norton's or Mace.

Ontrack—the same people who have sold several million copies of Disk Manager for hard disks—also has a utility program called DOSUTILS that provides tools to display and modify any physical sector of a hard disk, to scan for bad sectors, and to diagnose and analyze the disk. Steve Gibson's SpinRite, Prime Solution's Disk Technician, and Gazelle's OPTune are excellent hard disk tools for low-level formatting, defragmenting, and detecting potential bad sectors on a hard disk.

Norton Utilities

Norton is a program that everyone should have. Norton also has Norton Commander—a shell program—as well as the Norton Backup, a very good hard disk backup program. The Norton Company has recently merged with Symantec.

For more information, call Norton Company (213) 453-2361

Mace Utilities

Mace Utilities was developed by Paul Mace and was recently acquired by Fifth Generation Systems, the people who developed FastBack—the leading backup program.

For more information, call Fifth Generation Systems at (504) 291-7221.

PC Tools

PC Tools is an excellent program that just about does everything. It has data recovery utilities, hard disk backup, a DOS shell, a disk manager, and more.

For more information, call Central Point Software at (503) 690-8090.

SpinRite II

This software can check the interleave and reset it for the optimum factor without destroying your data. It can also test a hard drive and detect any marginal areas. SpinRite can maximize hard disk performance and prevent hard disk problems before they happen. Steve Gibson, the developer of SpinRite, writes a very interesting weekly column for InfoWorld Magazine.

For more information, call Gibson Research at (714) 830-2200.

Disk Technician

Disk Technician does essentially the same thing that SpinRite does, as well as a bit more. It has several automatic features and can now detect most viruses.

For more information, call Prime Solutions at (619) 274-5000.

OPTune

Optune is another utility that maximizes hard disk performance and is similar to SpinRite and Disk Technician. Gazelle Systems also developed both QDOS 3 (an excellent shell program) and Back-It 4 (a very good hard disk backup program).

For more information, call Gazelle Systems at (800) 233-0383.

CheckIt

CheckIt quickly checks and reports on your computers configuration, the CPU type, the memory amount, the installed drives, and the peripherals. It runs diagnostic tests of the installed items and can do performance benchmark tests.

For more information, call TouchStone Software Corp. at (213) 598-7746.

SideKick Plus

SideKick is in a class by itself. It was first released in 1984 and has been the most popular pop-up program ever since. It has recently been revised and enlarged so that it does much more than the simple calculator, notepad, calendar, and other utilities it had originally. It now has all of the original utilities plus scientific, programmer and business calculators, an automatic phone dialer, a sophisticated script language, and much more. SideKick loads into memory and pops up whenever you need it, no matter what program you happen to be running at the time.

For more information, call Borland International at (408) 438-8400.

Directory and disk management programs

Dozens of disk management programs help you keep track of your files and data on the hard disk and find, rename, view, sort, copy, and delete them—as well as many other useful things. These management programs can save an enormous amount of time and make your life much simpler.

XTreePro Gold

XTree was one of the first and still one of the best disk management programs. It has recently been revised and is now much faster and has several new features.

For more information, call Executive Systems at (800) 634-5545.

QDOS 3

QDOS 3 is a disk management program similar to XTree. It doesn't have quite as many features as XTree but is less expensive.

For more information, call Gazelle Systems at (800) 233-0383.

Tree86 3.0

Tree86 is another low-cost disk management program resembling XTree.

For more information, call Aldridge Company at (713) 953-1940

Wonder Plus 3.08

Wonder, or 1DIR, was one of the early disk management shells and has recently been revised and updated.

For more information, call Bourbaki, Inc., at (208) 342-5849.

Search utilities

I have about 3000 files on my hard disk in several subdirectories, and you can only imagine how difficult it is to keep track of all of them. I sometimes forget in which subdirectory I filed something, which could really cause some major difficulties. However, a couple of programs can go through all of my directories and look for a file by name.

Still, because you are only allowed eight characters for a filename, I have trouble remembering what's in each file. Luckily, several other programs can search through all the files and find almost anything that I want. I don't even have to know the full name of what I'm looking for; the programs will accept wildcards and tell me when things match.

Magellan 2.0
Magellan 2.0 is a very sophisticated program that can navigate and do global searches through files and across directories. It finds text and lets you view it in a window, as well as compresses files, does backup, compares, Undeletes, and several other excellent options.

For more information, call

Lotus
55 Cambridge Parkway
Cambridge, MA 02142
(800) 223-1662.

Several other search programs are not quite as sophisticated as Magellan but still work well (such as Gofer, from Microlytics). I listed some others that run under Windows in Chapter 13. In addition, you can also get some public domain and shareware search programs.

Computer Aided Design (CAD)
Several other companies besides those I mention offer CAD software. Check the computer magazines for their names and products.

AutoCAD
AutoCAD is a high-end, high-cost design program. It is quite complex with an abundance of capabilities and functions but is also rather expensive at about $3000 (some modules cost much less.) Still, AutoCAD is the IBM of the CAD world and has more or less established the standard for the many clones that have followed.

For more information, call Autodesk, Inc., at (415) 332-2344.

DesignCAD 2D and DesignCAD 3D
These CAD programs will do just about everything that AutoCAD will do at about one tenth of the cost. DesignCAD 3D allows you to make three-dimensional drawings. The list price for 2D is $299, while 3D is listed at $399. The discount houses are advertising them for as little as $142 and $188, respectively.

Tax programs
Because you have a computer, you might not have to pay a tax preparer to do your taxes; several tax programs can do the job for you. Unless you have a very compli-

cated income, you can figure it out quickly and easily. In many cases, the cost of the program would probably be less than the cost of having a tax preparer do your taxes.

Besides doing your own taxes, most of these programs will allow you to set up files and do the taxes of others. Of course, the software vendors would like to have each person buy a separate copy of the program, especially if you're in the tax preparation business. Many of them offer programs for professional tax businesses but usually at a much higher price.

All of the programs operate much like a spreadsheet in that the forms, schedules, and worksheets are linked together. When you enter data at one place, other affected data is automatically updated. Some of the programs are simply templates for Lotus 1-2-3 or Symphony and requires those programs to run, and most have a built-in calculator allowing you to do calculations before entering figures. Many of them allow "what-if" calculations to show you what your return would look like with various inputs, and some offer modules for some of the larger states such as New York and California. Most of the tax programs will allow you to print out IRS acceptable forms.

Andrew Tobias TaxCut

This program will handle most of the average returns and can be interfaced with Andrew Tobias' Managing Your Money, which is an excellent personal financial program. It does not offer state modules.

For more information, call Meca Ventures at (203) 222-9150.

J.K. Lasser's Your Income Tax

This program has a scratch pad, calculator, and next year tax planner. The popular J.K. Lasser's Tax Guide is included with the package.

For more information, call J.K. Lasser at either (800) 624-0023 or (800) 624-0024, in New York.

SwifTax

SwifTax has memo pads, calculator, context-sensitive help, and allows "what-if" tests. It has no state modules.

For more information, call Timeworks at (312) 948-9202.

TaxView

TaxView costs $119, with annual upgrades for only $55. California and New York modules cost $65 each.

TaxView is the PC version of Macintax, the foremost tax program for the Macintosh. It runs under Windows, (a run-time version is included) and is recommended to be used with a mouse. TaxView is very easy to learn and use, having a calculator, allowing "what-if" projections, and supporting a large number of IRS forms.

For more information, call SoftView, Inc., at (805) 388-2626. To reach Customer Service, dial (800) 622-6829.

TurboTax

TurboTax costs $75, with annual upgrades of $37.50. It is unique in that it offers modules for 41 states at $40 each. It also has an excellent manual and is fairly easy to install and learn. TurboTax starts out with a personal interview about your financial situation for the past year and then lists forms that you might need. Based on the present years taxes, it can estimate what your taxes will be for next year.

For more information, call ChipSoft, Inc., at (619) 453-8722.

Miscellaneous

Many programs are available dealing with such topics as accounting, statistics, finance, and others. Some programs are very expensive, but some are very reasonable.

Money Counts

Money Counts is a very inexpensive program that can be used at home or in a small business. With it, you can set up a budget, keep track of all of your expenses, balance your checkbook, plus do other things. It costs $40.

For more information, call Parsons Technology at (800) 223-6925.

It's Legal

It's Legal helps you create wills, leases, promissory notes, and other legal documents.

For more information, call Parsons Technology at (800) 223-6925.

WillMaker 4.0

WillMaker 4.0 is a low-cost program that can help you create a will, which is an important document. Everyone should have a will, no matter what the age or how much owned. Many people put it off because they don't want to take the time or don't want to pay the large lawyer fee. This inexpensive software can help you create a legal will.

For more information, call Nolo Press at (415) 549-1976.

Random House Encyclopedia

This software is basically the whole encyclopedia on disk. Find any subject very quickly.

For more information, call Microlytics at (716) 248-9150.

ACT!

ACT! is a program that lets you keep track of business contacts, schedules, business expenses, write reports, and about 30 other features.

    For more information, call Software at (800) 228-9228.

Form Express

Most businesses have dozens of forms that must be filled out and then, quite often, transferred into a computer. Form Express lets you easily design and fill in almost any kind of form on a hard disk. If necessary, the information can then be printed out.

    For more information, call Forms Express at (415) 382-6600.

Summary

I can't possibly mention all of the fantastic software available. Thousands of ready-made software programs will allow you to do almost anything with your computer. Look through any computer magazine for the reviews and ads, and you should be able to find programs for almost any application.

Chapter 15

Mail order and magazines

I've tried to list a few vendors when I mentioned a product, but I can't possibly list all of the sellers and sources for the products that you might need. (Face it, there's thousands of vendors and many more thousands of products.)

One of the best ways to find what you need is by looking through magazine advertising. The magazines also usually have a few informative articles scattered among the ads.

The life blood of magazines is advertising. Most magazines don't make enough money from their subscription rates to pay for the postage to mail them, so they depend on their advertisers to pay the rent, editors, and staff. (Then, if anything's left over, they give a few nickels to their writers.)

If you live near a large city, you can also visit the computer stores in town. You can actually see and touch the merchandise, and perhaps even try it out before you buy it.

Again, if you live near a large city, you'll also hear about a few swap meets every now and then. In the San Francisco Bay area and in the Los Angeles area, a swap meet is held almost every weekend.

Going to a swap meet is usually better than going to a computer store. You can look at price, and compare lots of items. Often several booths or vendors will all be selling the same thing that you need. Thus, you can take a pad and pencil, go to each vendor, get the best price, and then make your best deal.

Sometimes you can even haggle a bit with the vendors, because some will try to meet the price of their competition. The best time to haggle is near closing time: some of the vendors would rather sell their wares at a reduced price rather than pack them up and take them back to their store.

Mail order

Shopping at a local computer store can present a bit of a problem. To do it, you usually have to give up a bit of time, risk your life on the highway, and fight the traffic to get to the store. Then comes the most difficult part of the whole process—finding a parking space within a mile of the business.

In contrast, you can look through a magazine, find an ad for what you need, pick up your phone, order the components, and have them delivered to your door. As an extra plus, you might even pay 40% less than what you would pay at the local store.

Of course, what would life be without a few negatives? If you do order through the mail, you could have a wait of three or four weeks before you get your goodies. (Of course, you can usually pay extra and have them shipped by Federal Express or by UPS, in which case you could probably get them over night or within a couple of days—for that extra price).

Another negative point is that you are buying the components sight unseen. You have only the word of the advertiser that s/he will send them. However, if you use a bit of common sense and follow a few basic rules, you shouldn't have to worry.

One case not long ago, very expensive items were advertised for a very low price, but only a PO Box number was given as the place to which to send the money. Many people eagerly sent in money, thinking they had a deal too good to be true—and, of course, it was (unfortunately). When some of the people complained after much time passed, the Post Office investigated the box; by then, though, the operators were long gone.

This incident made the legitimate advertisers a bit unhappy because it made them all suspect; they worried that many people would no longer buy from them. Also, the computer magazine publishers became worried because, if people didn't buy from the advertisers, they would stop advertising. Then, without advertising revenue, the magazines would go belly-up in no time. In addition, it worried the Post Office and the Federal Trade Commission.

In response to the situation, the publishers and the advertisers got together and formed the Microcomputer Marketing Council (MMC) of the Direct Marketing Association.

Ten rules for ordering by mail

Advertisers are now policed fairly closely. Just to be on the safe side, though, you should follow the following rules when ordering through the mail:

- Make sure that the advertiser has a street address. In some ads, only a phone number is given. If you decide to buy from such a vendor, call and verify that there is a live person on the other end with a street number. Before you send any money, do a bit more investigation. If possible, look through past issues of the same magazine for previous ads. If the company has been advertising for several months, then it's probably okay.
- Check through the magazines for other vendors prices for this product; the prices should be fairly close. If the price appears to be a bargain too good to be true, then—well, you know the rest.
- Buy from a vendor who is a member of the Microcomputer Marketing Council (MMC) of the Direct Marketing Association (DMA) or some other recognized association. Different marketing associations now have about 10,000 members, all who have agreed to abide by the ethical guidelines and rules of the associations. Except for friendly persuasion and the threat of expulsion, the associations have little power over the members, but most of them realize the enormous stakes and put a great value on their membership. Thus, most who advertise in the major computer magazines are members.

    The Post Office, the Federal Trade Commission, the magazines, and the legitimate businessmen who advertise have taken steps to try to stop the fraud and scams.
- Do your homework. Know exactly what you want, state precisely the model, make, size, component, and any other pertinent information. Tell them which ad you are ordering from, ask them if the price is the same, if the item

is in stock, and when you can expect delivery. If the item is not in stock, indicate whether you will accept a substitute or want your money refunded. Ask for an invoice or order number. Ask for the person's name. Write down all of the information, the time, the date, the company's address and phone number, description of item, and promised delivery date. Save any and all correspondence.

- Ask if the advertised item comes with all the necessary cables, parts, accessories, software, etc. Ask what the warranties are. Ask what the seller's return and refund policies are. Also find out with whom should you correspond if you have a problem.
- Don't send cash—you will have no record of it. If possible, use a credit card. Then, if you do have a problem, you can possibly have the bank refuse to pay the amount. A personal check might cause a delay of three to four weeks while the vendor waits for it to clear, while a money order or credit card order should be filled and shipped immediately. Keep a copy of the money order.
- If you have not received your order by the promised delivery date, notify the seller.
- Try the item out when you receive it. If you have a problem, notify the seller immediately, by phone and then in writing. Give all details. Don't return the merchandise unless the dealer gives you authorization. Make sure to keep a copy of the shipper's receipt or packing slip or evidence that it was returned.
- If you believe the product is defective or if you have a problem, reread your warranties and guarantees. Reread the manual and any documentation. You can easily make an error or misunderstand how an item operates if you are unfamiliar with it. Before you go to a lot of trouble, try to get some help from someone else. At least, get someone to verify that you do have a problem. Many times, a problem will disappear and the vendor will not be able to duplicate it.
- Try to work out your problem with the vendor. If you cannot, then write to the consumer complaint agency in the seller's state. You should also write to the magazine and to

    Direct Marketing Association (DMA)
    11 W. 42nd St.
    New York, NY 10036
    (212) 768-7277

Federal Trade Commission rules

Here is a brief summary of the FTC rules:

- The seller must ship your order within 30 days unless the ad clearly states that it will take longer.

- If it appears that the seller cannot ship when promised, he must notify you and give a new date. He must give you the opportunity to cancel the order and refund your money if you so desire.
- If the seller notifies you that he cannot fill your order on time, he must include a stamped self-addressed envelope or card so that you can respond to his notice. If you do not respond, he can assume that you agree to the delay. He still must ship within 30 days of the end of the original 30 days or cancel your order and refund your money.
- Even if you consent to a delay, you still have the right to cancel at any time.
- If you cancel an order that has been paid for by check or money order, he must refund the money. If you paid by credit card, your account must be credited within one billing cycle. Store credits or vouchers in place of a refund are not acceptable.
- If the item you ordered is not available, the seller cannot send you a substitute without your express consent.

You should try by all means to work out your problems with the vendor. If it looks hopeless, though, then contact the DMA at (212) 768-7277, your local US Postal Inspector, your local Better Business Bureau, your State Consumer Affairs, or the Consumer Protection Agency. You can also call the Federal Trade Commission at (202) 768-3768 to complain if you can't resolve your problem.

Computer magazines

The magazine business is highly competitive. Many enter the business, but many don't survive. Here's a current list of some magazines that you should subscribe to if you want to keep up:

Byte Magazine
P.O. Box 558
Hightstown, NJ 08520-9409

Compute!
P.O. Box 3244
Harlan, IA 51593-2424

Computer Currents
5720 Hollis St.
Emeryville, CA 94608

Computer Monthly
P.O. Box 7062
Atlanta, GA 30357-0062

Computer Graphics World
P.O. Box 122
Tulsa, OK 74101-9966

Computer Shopper
P.O. Box 51020
Boulder, CO 80321-1020

Data Based Advisor
P.O. Box 3735
Escondido, CA 92025-9895

Home Office Computing
P.O. Box 51344
Boulder, CO 80321-1344

LAN Magazine
Miller Freeman Publications
P.O. Box 50047
Boulder, CO 80321-0047

MicroTimes Magazine
5951 Canning St.
Oakland, CA 94609

PC Computing
P.O. Box 50253
Boulder, CO 80321-0253

PC World Magazine
P.O. Box 51833
Boulder, CO 80321-1833

PC Magazine
P.O. Box 51524
Boulder, CO 80321-1524

PC Today
P.O. Box 85380
Lincoln, NE 68501-9815

Personal Workstation
P.O. Box 51615
Boulder, CO 80321-1615

Publish!
P.O. Box 51966
Boulder, CO 80321-1966

Unix World
P.O. Box 1929
Marion, OH 43306

Free magazines to qualified subscribers

The magazines listed next are free to qualified subscribers. The subscription price of a magazine usually does not come anywhere near covering the costs of publication, mailing, distribution, and other costs; most magazines depend almost entirely on advertisers for their existence. Thus, the more subscribers that a magazine has, the more it can charge for its ads. Naturally they can attract a lot more subscribers if the magazine is free.

Computer Buying World, PC Week and InfoWorld are excellent magazines, so popular in fact that the publishers must limit the number of subscribers— they can't possibly accommodate all the people who have applied. They've set standards that must be met in order to qualify. Still, they don't publish the standards, so even if you answer all of the questions on the application, you still might not qualify.

To get a free subscription, you must write to the magazine for a qualifying application form. The form will ask several questions such as how you are involved with computers, the company you work for, whether you have any influence in purchasing the computer products listed in the magazines and several other questions that gives them a very good profile of their readers.

I once filled out a qualifying form for a free magazine and waited for awhile, but I never received the magazine. Later, I met one of the editors at a computer show and complained. He said "Well, it's probably your own fault. You just didn't lie enough on the form."

Those were his words. I would never encourage you to lie, but it might help you qualify for a free subscription if you exaggerate just a bit here and there (especially when it asks what your responsibilities are for the purchasing of computer equipment). I'm pretty sure that they won't send the FBI after you to verify your answers.

FaxBack

For years, many magazines have printed a number on each ad and then included a post card with all of the ad numbers in the back of the magazine. If a reader wanted more information about a certain ad, he circled the ad number on the "bingo" card and sent it to the magazine. Sometimes one had to wait four or five weeks to get a response.

Computer Buying World is a new magazine that's instituted a unique FaxBack system. They number each of their ads, and if a person wants more information about a product advertised in the magazine, they can call (617) 246-5089. From a touchtone phone, the number of the product ad is typed in and more information will be sent back to your FAX machine immediately.

This list of magazines is not nearly complete; hundreds of trade magazines are sent free to qualified subscribers (Cahners Company alone publishes 32 different trade magazines). Many of the trade magazines are highly technical and narrowly specialized.

Computer Buying World
P.O. Box 3020
Northbrook, IL 60065-9847

PC Week
P.O. Box 5920
Cherry Hill, NJ 08034

InfoWorld
1060 Marsh Rd.
Menlo Park, CA 94025

Computer Design
Circulation Dept.
P.O. Box 3466
Tulsa, OK 74101-3466

Computer Systems News
600 Community Dr.
Manhasset, NY 11030

Communications Week
P.O. Box 2070
Manhasset, NY 11030

Computer Reseller News
P.O. Box 2040
Manhasset, NY 11030

Computer Products
P.O. Box 14000
Dover, NJ 07801-9990

Computer Technology Review
924 Westwood Blvd., Suite 650
Los Angeles, CA 90024-2910

California Business
Subscription Dept.
P.O. Box 70735
Pasadena, CA 91117-9947

Designfax
P.O. Box 1151
Skokie, IL 60076-9917

Discount Merchandiser
215 Lexington Ave.
New York, NY 10157

EE Product News
P.O. Box 12982
Overland Park, KS 66212-9817

Electronics
A Penton Publication
P.O. Box 985061
Cleveland, OH 44198-5061

Electronic Manufacturing
Lake Publishing
P.O. Box 159
Libertyville, IL 60048-9989

Electronic Publishing & Printing
650 S. Clark St.
Chicago, IL 60605-9960

Federal Computer Week
P.O. Box 602
Winchester, MA 01890-9948

Identification Journal
2640 N. Halsted St.
Chicago, IL 60614-9962

ID Systems
174 Concord St.
P.O. Box 874
Peterborough, NH 03458-0874

Automatic I.D. News
P.O. Box 6170
Duluth, MN 55806-9870

Lan Times
122 East, 1700 South
Provo, UT 84606

Lasers & Optronics
301 Gibraltar Dr.
P.O. Box 601
Morris Plains, NJ 07950-9827

Machine Design
Penton Publishing
P.O. Box 985015
Cleveland, OH 44198-5015

Modern Office Technology
Penton Publishing
1100 Superior Ave.
Cleveland, OH 44197

Manufacturing Systems
P.O. Box 3008
Wheaton, IL 60189-9972

Medical Equipment Designer
Huebcore Communications
29100 Aurora Rd., #200
Cleveland, OH 44139

Mini-Micro Systems
P.O. Box 5051
Denver, CO 80217-9872

Modern Office Technology
1100 Superior Ave.
Cleveland, OH 44197-8032

Office Systems 90
P.O. Box 3116
Woburn, MA 01888-9878

Office Systems Dealer 90
P.O. Box 2281
Woburn, MA 01888-9873

Photo Business
1515 Broadway
New York, NY 10036

The Programmer's Shop
5 Pond Park Rd.
Hingham, MA 02043-9845

Quality
P.O. Box 3002
Wheaton, IL 60189-9929

Reseller Management
301 Gibraltar
Box 601
Morris Plains, NJ 07950-9811

Robotics World
6255 Barfield Rd.
Atlanta, GA 30328-9988

Unix Review
Circulation Dept.
P.O. Box 7439
San Francisco, CA 94120-7439

Scientific Computing & Automation
301 Gibraltar Dr.
Morris Plains, NJ 07950-0608

Surface Mount Technology
Lake Publishing Corp.
P.O. Box 159
Libertyville, IL 600048-9989

Public domain software

The next short list shows companies that provide public domain, shareware, and low-cost software.

PC-SIG
1030D E. Duane Ave.
Sunnyvale, CA 94086
(800) 245-6717

MicroCom Systems
3673 Enochs St.
Santa Clara, CA 95051
(408) 737-9000

Public Brand Software
P.O. Box 51315
Indianapolis, IN 46251
(800) 426-3475

Software Express/Direct
P.O. Box 2288
Merrifield, VA 22116
(800) 331-8192

Selective Software
903 Pacific Ave. Suite 301
Santa Cruz, CA 95060
(800) 423-3556

The Computer Room
P.O. Box 1596
Gordonsville, VA 22942
(703) 832-3341

Softwarehouse
3080 Olcott Dr., Suite 125A
Santa Clara, CA 95054
(408) 748-0461

PC Plus Consulting
14536 Roscoe Blvd., #201
Panorama City, CA 91402
(818) 891-7930

Micro Star
P.O. Box 4078
Leucadia, CA 92024-0996
(800) 443-6103

International Software Library
511 Encinitas Blvd., Suite 104
Encinatas, CA 92024
(800) 992-1992

National PD Library
1533 Avohill
Vista, CA 92083
(619) 941-0925

Computers International
P.O. Box 6085
Oceanside, CA 92056
(619) 630-0055

Shareware Express
27601 Forbes Rd., #37
Laguna Niguel, CA 92677
(714) 367-0080

Most of the companies listed here can provide a catalog listing of their software, for which some charge a small fee. Write to them or call them for details and latest prices.

This list is not complete; you might find several other companies advertised in some of the magazines listed earlier.

Mail-order books

One of the better ways to learn about computers is through books. Many bookstores will ship computer books to you. One that carries all of my books is

CompuBooks
728 B St.
Coeur d'Alene, ID 83814
(800) 765-1714

Also, several companies publish computer books. One of these companies is

TAB Books (a division of McGraw-Hill)
Blue Ridge Summit, PA 17294-0850
(800) 822-8138.

Call or write to them for a catalogue listing of the many books that they publish. I highly recommend them. Of course, the fact that they published the book that you're now holding in your hands doesn't mean that I'm biased (well, maybe just a little bit).

Chapter 16
Troubleshooting

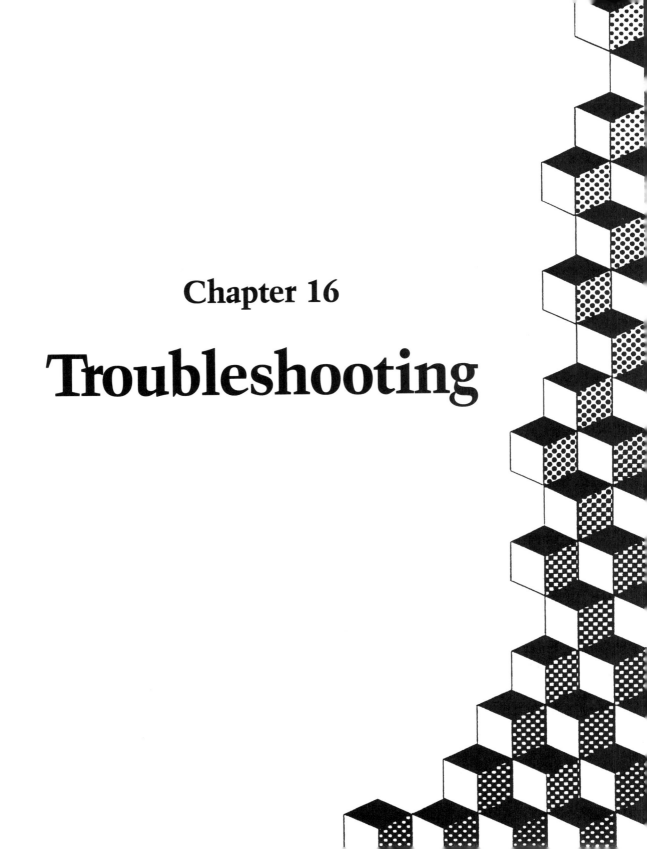

If you assembled your computer properly, it should work perfectly. However, there is always the possibility that something was not plugged in correctly or that some minor error was made.

I have a friend who works for a large computer mail-order firm, checking and repairing all of the components returned from the customers. I asked him what the biggest problem was. He answered "People just don't read and follow the instructions, or they make errors and don't check their work."

By far, then, the greatest problem in assembling a unit or adding something to a computer is not following the instructions. Quite often, the person not following the instructions isn't at fault. Although I've worked in the electronic industry for over 30 years, sometimes I have great difficulty in trying to decipher and follow the instructions in some manuals. Sometimes a very critical instruction or piece of information is buried inconspicuously in the center of a 450-page manual, and sometimes the manual is just poorly written.

If you have just assembled or added something to your computer, turn it on and check it out before you put the cover on. If something's wrong, this will usually make it easier to find the problem.

Before you turn the computer on, though, recheck all the cables and any boards or chips. Make sure that they are seated properly and in the right place. Read the instructions again and then turn on the power. If it works, put the cover on and button it up.

If you have added a board or some accessory and your computer doesn't work, remove the item and try the computer again.

Fewer bugs today

In the early days, many bugs and errors surfaced in clone computers. The Far East manufacturers didn't spend a lot of money on Quality Control and testing. However, most computer manufacturers have been making the parts long enough now that the designs have been firmed up and most bugs eliminated.

Document the problem and write it down

Chances are, if your computer is going to break down, it will do it at the most inopportune time (this is one of the basic tenets of Murphy's laws).

If the computer breaks down, try not to panic. Ranting, cussing, and crying might make you feel better but won't solve the problem. Under no circumstances should you beat on your computer with a chair or a baseball bat.

Instead, get out a pad and pencil and write down everything as it happens (otherwise you might forget). Write down all the particulars, how the cables were plugged in, the software that you were running, and anything that might be pertinent. You might get error messages on your screen; if you do, use the PrtScr (Print Screen) key to print out the messages.

If you can't solve the problem, you might have to call a friend or your vendor for help. If you have all the written information before you, it will help. Try to call from your computer, if possible, while it's acting up.

Levels of troubleshooting

Troubleshooting has many different levels. Advanced troubleshooting requires much sophisticated equipment such as oscilloscopes, digital meters, logic probes, signal generators, and lots of training. Most problems that you encounter, however, will be rather minor, so you won't need all that equipment and training. Most of your problems can be solved with just a little common sense and the use of the five senses of sight, sound, touch, smell, and taste (well, actually, you probably won't be using taste very often).

Electricity—the life blood of the computer

Troubleshooting will be a little easier if you know just a little of the electronic basics.

Computers are possible because of electricity. An electric charge is formed when there's an imbalance or excess amount of electrons at one pole. The excess electrons will flow through whatever path they can find to get to ground or to the other pole (similar to water flowing downhill to find its level).

Most electric or electronic paths have varying amounts of resistance so that work or heat is created when the electrons pass through them. For instance, if a flashlight is turned on, electrons will pass through the bulb, which has a resistive element. The heat generated by the electrons passing through the bulb will cause it to glow red-hot and create light. If the light is left on for a period of time, all of the excess electrons from the positive anode of the battery will pass through the bulb to the negative pole of the battery. At this time, the amount of electrons at the negative and positive poles will be the same. A perfect balance will exist and the battery will be dead.

The computer is made up of circuits and boards that have transistors, resistors, capacitors, inductors, motors, and many other components, all of which perform a useful function when electricity passes through them. Out of all the components, however, the transistor is the most important. It can amplify the micro-voltage generated by the data on a disk drive, divide and shunt the voltages to perform the desired function, store data in RAM, read the data, control the disk drives, monitor, and printer, and do a thousand other things necessary for the computer to operate.

Occasionally, too many electrons might find their way through a weakened component and burn it out; or, for some reason, the electrons might be shunted through a different path, possibly causing an intermittent, partial, or complete failure.

The basic components of a computer

The early IBM PC had an 8088 CPU, a BIOS chip, and four other basic support chips—the 8259 Interrupt Controller, the 8237 DMA Controller, the 8253/8254 Programmable Interval Timer, and the 8255 Programmable Input/Output Controller. These same chips are found in the 8086, the 80286, the 80386SX, the 80386DX, the 80486SX, and 89486DX. You will find two DMA and two Interrupt Controllers in the 286, 386, and 486. You might not be able to see these chips on some motherboards, especially the baby motherboards, because they are usually integrated into a large VLSI package.

The CPU is the brains of the computer, controlling the basic operation by sending and receiving control signals and memory addresses. It sends and receives data along the bus to and from other parts of the system, and it also carries out computations, numeric comparisons, and many other functions in response to software programs.

The Basic Input/Output System (BIOS) is second only to the CPU in importance. Every time you turn your computer on, the BIOS does a Power On Self Test (POST) of all the major components. BIOS is like a dedicated boss in charge of a factory; s/he comes in early, checks all of the equipment, and then opens for business. If something is wrong with one of the major components, s/he might not open for business.

The BIOS boss is in charge of all of the mundane business activity. It might be inundated with requests for attention to various jobs; it must determine whether to interrupt a job to fulfill a new request.

If you have an older machine, your BIOS might be unaware of many of the newer products. I have an old 1984 AT 286 on which I could not use a 1.44M floppy because the BIOS simply didn't know that this type of component existed. I replaced the BIOS with a new set of chips and can now easily run the 1.44M floppy as well as several other modern functions.

The 8259 Programmable Interrupt Controllers respond to interrupt requests generated by system hardware components. These requests could be from such components as the keyboard, disk drive controller, and system timer.

The 8237 DMA Controllers are able to transfer data to and from the computer's memory without passing it through the CPU, which allows I/O from the disk drives without CPU involvement.

The Programmable Interval Timer 8253 or 8254 generates timing signals for various system tasks.

The Input/Output Controller 8255 provides an interface between the CPU and the I/O devices.

Of course, several other chips are interrelated to each of these main chips, and all of the main chips are interrelated. Because they are all so intimately interrelated, a failure in any main or minor chip can cause the whole circuit to fail. The actual defect can be very difficult to pinpoint because of the interrelated chips.

Power On Self Test (POST)

Every time you turn your computer on, the BIOS does a Power On Self Test or POST. It checks the RAM, the floppy drives, the hard disk drives, the monitor, the printer, the keyboard, and other peripherals that you have installed.

If it does not find a unit or if the unit is not functioning correctly, the computer will beep and display an error code (with codes starting at 100 and possibly going up to 2500). Ordinarily, the codes will not be displayed if no problem has occurred. If a problem occurred, the last two digits of the code will be something other than 00s. Each BIOS manufacturer develops their own codes, so some slight differences are noticeable. Some codes you might see are as follows:

Codes	Explanations
101	Motherboard failure.
109	Direct Memory Access test error.
121	Unexpected hardware interrupt occurred.
163	Time and date not set.
199	User indicated configuration not correct.
201	Memory test failure.
301	Keyboard test failure or a stuck key.
401	Monochrome display and/or adapter test failure.
432	Parallel printer not turned on.
501	Color Graphics display and/or adapter test failure.
601	Diskette drives and/or adapter test failure.
701	Math Coprocessor test error.
901	Parallel printer adapter test failure.
1101	Asynchronous Communications adapter test failure.
1301	Game control adapter test failure.
1301	Joystick test failure.
1401	Printer test failure.
1701	Fixed disk drive and/or adapter test failure.
2401	Enhanced Graphics display and/or adapter test failure.
2501	Enhanced Graphics display and/or adapter test failure.

DOS has several other error messages; you could see something like Invalid Drive, File Not Found, or Printer Not Ready. All of the 50+ DOS commands can generate error messages. The DOS 5.0 manual has 668 pages, but only two obscure error messages are listed in the 29-page index. In addition, many of the messages are not very clear; the DOS manual explains some of them, but you might have to dig through the whole 668-page manual to find what you're looking for.

The impossible quest

If you ask the computer to do something it can't do, often it just will go off into Never-Never Land trying to do it anyway. You can pound on the keyboard as much as you want, but the computer will ignore you and keep trying to fulfill your request. Often, your only way out is to turn off the power and reboot. Of course, if you happened to be working on a program in memory, the data will be lost when you reboot.

Power supply

Most of the components in your computer are fairly low power and low voltage. The only high voltage in your system is in the power supply and that's well enclosed, so you aren't in any danger of shock if you open your computer and put your hand inside it. However, you should *NEVER EVER* connect or disconnect a board or cable while the power is on; fragile semiconductors can be destroyed if you do.

Most of the power supplies have short-circuit protection. If too large a load is placed on them, they will drop out and shut down, similar to what happens when a circuit breaker is overloaded. Most of the power supplies are designed to operate only with a load. If you take one out of the system and turn it on without a load, most of them will not work. You can plug in a floppy drive to act as a load if you want to check the voltages out of the system.

The fan in the power supply should provide all the cooling normally needed. However, if you stuff your computer in the corner and pile things around it and shut off all its circulation, it could possibly overheat. Heat is an enemy of semiconductors, so give your computer plenty of breathing room.

The semiconductors in your computer have no moving parts. If they were designed properly, they should last indefinitely. As I just said, heat is an enemy and can cause semiconductor failure, although the fan in the power supply should provide adequate cooling. All of the openings on the back panel that correspond to the slots on the motherboard should have blank fillers, and even the holes on the bottom of the chassis should be covered with tape. Closing off all random apertures forces the fan to draw air in from the front of the computer, pull it over the boards, and exhaust it through the opening in the power supply case, thus cooling off the whole computer. Nothing should be placed in front of or behind the computer that would restrict this air flow.

If you don't hear the fan when you turn on your computer and you're sure it's not running, then the power supply could be defective.

The 8-bit slot connectors on the motherboard have 62 contacts—31 on the A side and 31 on the B side. The black ground wires connect to B1 of each of the slots. B3 and B29 have +5 VDC, B5 has −5 VDC, B7 has −12 VDC, and B9 has +12 VDC. These voltages go to the listed pins on each of the eight plug-in slots (see Table 16-1).

Table 16-1 Hooking up the pin connections and wire colors from the power supply.

Disk drive power supply connections

Pin	Color	Function
1	Yellow	+12 VDC
2	Black	Ground
3	Black	Ground
4	Red	+5 VDC

Power supply connections to the motherboard

P8

Pin	Color	Function
1	White	Power Good
2	No connection	
3	Yellow	+12 VDC
4	Brown	-12 VDC
5	Black	Ground
6	Black	Ground

P9

Pin	Color	Function
1	Black	Ground
2	Black	Ground
3	Blue	-5 VDC
4	Red	+5 VDC
5	Red	+5 VDC
6	Red	+5 VDC

Instruments and tools

For high levels of troubleshooting, a person would need some rather sophisticated and expensive instruments to do a thorough analysis of a system. Necessary would be a good high frequency oscilloscope, a digital analyzer, a logic probe, and several other expensive pieces of gear. You would also need a test bench with a power supply, disk drives, and a computer with some empty slots so that you could plug in suspect boards and test them.

You would also need a volt ohmmeter, some clip leads, a pair of side cutter dikes, a pair of long nose pliers, various screwdrivers, nut drivers, a soldering iron and solder, and lots of different size screws and nuts.

In addition, plenty of light over the bench and a flashlight would help you too, or at least a small light to brighten up dark places in the case.

Most importantly, you will need quite a lot of training and experience. For many problems, however, just a little common sense will tell you what is wrong.

Common problems

For most of the common problems, you won't need a lot of test gear; you can often solve them by using your eyes, ears, nose, and hands.

Using your eyes, you could see that a cable isn't properly plugged in, a board isn't completely seated, or a switch isn't correctly set (as well as other obvious things).

Using your ears, you can notice unusual sounds. The only sound coming from your computer should be the clack of your drive motors and the whirr of the fan in the power supply.

Using your nose, you can smell unusual scents. If you've ever smelled a burned resistor or a capacitor, you won't forget it. If you do smell something very unusual, try to locate where the smell is coming from.

Using your fingers, you can sense heat. If you touch the components and some of them seem unusually hot, you might have found the cause of your problem. Don't worry about getting shocked. Except for the insides of your power supply, no voltage in your computer should be above 12 volts, so you can safely touch the components.

Electrostatic discharge (ESD)

Before you touch or handle any of the components, you should ground yourself and discharge any static voltage that you might have built up. You can discharge yourself by touching an unpainted metal part of the case of a computer or other plugged-in device. You can build up a charge of 4000 or more electrostatic voltage. If you walk across some carpets and then touch a brass door knob, you can sometimes see a spark fly and often get a shock. Most electronic assembly lines have the workers wear a ground strap whenever they are working with any electrostatic discharge sensitive components.

Recommended tools

The following are some tools that you should have around the house, even if you never have any computer problems.

- Several sizes of screwdrivers. A couple of them should be magnetic for picking up and starting small screws. You can buy magnetic screwdrivers, or you can make one yourself. Just take a strong magnet and rub it on the blade of the screwdriver a few times (the magnets on cabinet doors or the voice coil magnet of a loudspeaker will do). Be very careful with any magnet around your floppy diskettes because the magnetic field can erase them.

- A small screwdriver with a bent tip that can be used to pry up ICs. Some of the larger ICs are very difficult to remove. One of the blank fillers for the slots on the back panel also makes a good prying tool.
- A couple pairs of pliers. You should have at least one pair of long nose pliers.
- A pair of side cutter dikes for clipping leads of components and cutting wire. You might buy a pair of cutters that also have wire strippers.
- A soldering iron. You shouldn't have to do any soldering, but you never know when you might need to. A soldering iron comes in handy around the house many times. Of course, get some solder too.
- No home should be without a volt ohmmeter, which can be used to check for the correct wiring in house wall sockets. (The wide slot should be ground). They can be used to check switches, wiring continuity in your car, house, stereo, phone lines, etc. Also, you could check for the proper voltages in your computer (only four— +12 volts, −12 volts, +5 volts, and −5 volts).

You can buy a relatively inexpensive volt ohmmeter at any of the Radio Shack or electronic stores.

- Several clip leads. You can buy them at the local Radio Shack or electronic store.
- A flashlight for looking into the dark places inside the computer.

How to find the problem

Follow these steps in order to find a problem:

1. If you suspect a board and you have a spare or can borrow one, swap it. Before removing any board or changing any switch or jumper setting, make a rough diagram; you can easily forget how it was originally set. You could make the problem even worse.
2. If you suspect a board but don't know which one, take the boards out to the barest minimum. Then add them back until the problem develops.
   Caution: Always turn off the power when plugging in or unplugging a board or cable!
3. Wiggle the boards and cables to see if it is an intermittent problem. Many times a wire can be broken and still make contact until it is moved.

   Check the ICs and connectors for bent pins. If you have installed memory ICs and get errors, check to make sure that they are seated properly and that all the pins are in the sockets. If you swap an IC, make a note of how it is oriented before removing it. There should be a small dot of white paint or a U-shaped indentation at the end that has pin 1. If you forgot to note the orientation, look at the other ICs; most of the boards are laid out so that all of the ICs are oriented the same way. The chrome fillers used to cover the unused slots in the back of the case make very good tools for prying up ICs.

You might also try unplugging a cable or a board and plugging it back in. Sometimes the pins might be slightly corroded or not seated properly. Before unplugging a cable, you might put a stripe on the connector and cable with a marking pen or nail polish so that you can easily see how they should be plugged back in.

The problem could be in a DIP switch. You might try turning it on and off a few times.

Caution: Again, always write down the positions before touching the switches.

Make a pencil mark before turning a knob or variable coil or capacitor so that it can be returned to the same setting when you find out that it didn't help. Better yet, resist the temptation to reset these types of components. Most were set up using highly sophisticated instruments and don't usually change enough to cause a problem.

4. If you are having monitor problems, check the switch settings on the motherboard. Several different motherboards exist; some have dip switches or shorting bars that must be set to configure the system for monochrome, CGA, EGA or VGA.

    Most monitors also have fuses. You might check them. Also check the cables for proper connections.

5. Printer problems, especially serial type, are so numerous that we will not even attempt to list them here. Many printers today have parallel and serial, with IBM defaulting to the parallel system. If at all possible, use the parallel port; parallel has very few problems compared to serial.

    Most printers have a self-test. It might run this test fine but then completely ignore any efforts to make it respond to the computer if the cables, parity, and baud rate are not properly set.

6. Sometimes the computer will hang up, possibly when you tell it to do something that it can't. You can usually warm boot your computer by pressing Ctrl-Alt-Del. Of course, the reboot would wipe out any file in memory that you might have been working on. Occasionally the computer will not respond to a warm boot. In that case, you will have to switch off the main power, let it sit for a few seconds, and then power up again. Always wait for the hard disk to wind down and stop before turning the power back on.

Diagnostic and utility software

When IBM came out with the XT, they developed a diagnostic or set-up disk included with every machine. It checked the keyboard, the disk drives, the monitor, peripherals, and performed several other tests. When the AT was released, the diagnostic disk was revised a bit to include even more tests. You had to have the disk to set the time, date, and all of the other on-board CMOS system configuration.

Most BIOS chips now have many of the diagnostic routines built-in. These routines allow you to set the time and date, tell the computer what type of hard drive and floppies are installed, the amount of memory, the wait states, and several other functions. The AMI and DTK BIOS chips have a very comprehensive set of built-in diagnostics. They can allow hard and floppy disk formatting, check speed of rotation of disk drives, do performance testing of hard drives, and several others.

I mentioned these utility software programs in Chapter 14. Many of them have a few diagnostics among the utilities.

- Norton Utilities. It also includes several diagnostic and test programs such as Disk Doctor, Disk Test, Format Recover, Directory Sort, and System Information.
- Mace Utilities. It does about everything that Norton does and a few other things. It has recover, defragment, diagnose, remedy, and other very useful programs primarily for the hard disk.
- PC TOOLS. It has several utilities much like the Norton and Mace Utilities. It has a utility that can recover data from a disk that has been erased or reformatted. It has several other data recovery and DOS utilities and can be used for hard disk backup. It also has several utilities such as those found in SideKick.
- SpinRite (from Gibson Research), Disk Technician (from Prime Solutions), Optune (from Gazelle Systems) and DOSUTILS (from Ontrack). These are utilities that allow you to diagnose, analyze and optimize your hard disk.
- CheckIt (from TouchStone Software). It checks and reports on your computer configuration by letting "you look inside your PC without taking off the cover." It reports on the type of processor, amount of memory, video adapter, hard and floppy drives, clock/calendar, ports, keyboard, and mouse (if present). It also tests the motherboard, hard and floppy disks, RAM, Ports, Keyboard, mouse, and joystick. It can also run a few Benchmark speed tests.
- Xidex Corporation, (408) 988-3472. They manufacture some of the better floppy diskettes. Dysan, one of their branches, has developed the Interrogator software, which can check a floppy disk drive for head alignment and performance and do several other diagnostic tests. If you are having trouble reading software on a certain drive, a quick test with this software will tell you whether it is the drive or the software.

What to do if it is completely dead

Software diagnostics are great in many cases, but if the computer is completely dead, the software won't do you any good.

If the computer is entirely dead, you should first check the power. If you don't have a voltmeter, plug a lamp in the same socket and see if it lights. Check your power cord. Check the switch on the computer. Check the fan in the power supply—is it turning? Check the monitor, power cord, fuses, and adapter.

If these all seem to be okay, then you might have some serious problems. You probably need some high-level troubleshooting.

Software problems

I have had far more trouble with software than I have had with hardware. Quite often, I'm at fault for not taking the time to completely read the manuals and instructions. For instance, I tried to install Charisma, which is a large program and works under Windows. I kept getting errors, and it wouldn't load. I finally read the manual and found that Charisma requires at least 500K of memory to install. I have several drivers, TSRs, and other things in my CONFIG.SYS file that eat up a lot of RAM. Thus, I had to boot up with a "plain vanilla" diskette that had a very simple CONFIG.SYS file that left me over 500K of RAM. I had no trouble after that.

DR DOS 6.0 and MS-DOS 5.0 can load drivers, TSRs, and other things into memory above 640K, leaving as much as 630K of free RAM. Windows 3.0 comes with a HIMEM.SYS that is necessary for accessing extended memory and is loaded with the CONFIG.SYS file. If HIMEM.SYS is loaded, it will conflict with the DR DOS high-memory files. You can run one or the other; but if both are loaded, the computer won't boot up.

You'll probably run into thousands of other software problems. Many vendors have support programs for their products; if something goes wrong, you can call them. A few of them even offer toll-free numbers, although most make you pay for the call. Some companies charge for their support, and some installed a 900 telephone number. You are charged a certain amount for the amount of time on the phone. It can cost a lot of money to maintain a support staff.

If you have a software problem, write down everything that happens. Before you call, try to duplicate the problem or make it happen again. Carefully read the manual. When you call, you should be at your computer, with it turned on and with the problem on the screen if possible.

Also, before you call, have the serial number of your program handy. One of the first things they will probably ask is for your name and serial number. If you have bought and registered the program, it will be in their computer.

Compatibility problems still occur in areas of updates and new releases of software. I had lots of problems trying to get the latest WordStar release to work with files I had created with an early version. It was mostly my fault because I didn't take the time to read the manual. Like so many other updates and revisions, they make them bigger and better; but out of necessity, they often change the way things were done earlier.

Thousands of software programs are available, and most of them are reasonably bug-free. Still, millions of things can go wrong if you don't follow the exact instructions and procedures. Of course, in many cases, the exact instructions and procedures are not very explicit.

I can't possibly list all of the possible software problems.

Hardware problems

Many things can go wrong with hardware also. Some things might happen once but never again. For instance, a 3.5″ drive had been working perfectly. One day, I tried to load a program and it gave me the DOS error message "Sector not found, error reading drive B:." I thought that maybe I had screwed up my program diskette, so I tried it on one of my other computers and it worked fine. I pulled the drive out and took the cover off, hooked it up, and tried to read a diskette, but I could see that the head never moved.

The head is attached to a long screw-type worm gear. The head actuator motor turns the screw and moves the head to the various tracks. I turned the power off and moved the head manually. It seemed to be stuck at first but then moved freely. I turned the power back on and then it worked perfectly. I have not had any trouble since then, and it will probably never happen again.

Likewise, I ported my 486 off to a User Group meeting, took it all apart, and demonstrated how easy it was to assemble. When I got back home, neither of my floppy drives would operate. Because neither drive would work, I figured that the problem could be the cable or the floppy controller portion of my hard disk controller. Thus, I installed a new cable, but it didn't help. I then plugged in a new floppy controller and still could not read or write to the floppies. At this point, I really began to worry; maybe I had damaged my $4450 motherboard when I transported it to the meeting.

I couldn't understand how both floppy drives could be bad, but I disconnected them and plugged in another 1.2M drive. This time it worked perfectly. I then reinstalled the 1.44M drive and it also worked. Apparently, whatever was wrong with the 1.2M A drive also kept the B drive from operating; I had evidently damaged it in handling. I was unhappy and disgusted with myself for being so careless, but I was also happy that it was only a $60 floppy drive and not my very expensive motherboard. Because the drive would cost more to troubleshoot and repair than it was worth, I scrapped it and bought a new one.

Computers are dumb and very unforgiving. You can easily plug a cable into a disk drive backwards or forget to set a switch. Sometimes you have trouble determining if a hardware problem is caused by software or vice versa. You can't possibly address every problem, but you can still use common sense to solve your dilemmas. One of the best ways to find answers is to ask someone who has had the same problem, which you can do most easily at a Users Group. If at all possible, join one.

You can also get help from local Bulletin Boards. Your computer is not complete without a modem that allows you to contact them.

Several local computer magazines list User Groups and Bulletin Boards as a service to their readers. The nationally published Computer Shopper prints a very comprehensive list each month.

Thousands of things can go wrong. Often, only one way exists to do something right but ten thousand ways exist to do it wrong.

Is it worth repairing?

If you find a problem on a board, disk drive, or some component, you might try to find out what it would cost before having it repaired. With the low-cost clone hardware available, it is often less expensive to scrap a defective part and buy a new one.

Again, if at all possible, join a Users Group and become friendly with all of the members. They can be one of your best sources of troubleshooting. Most of them have had similar problems and are glad to help.

Glossary

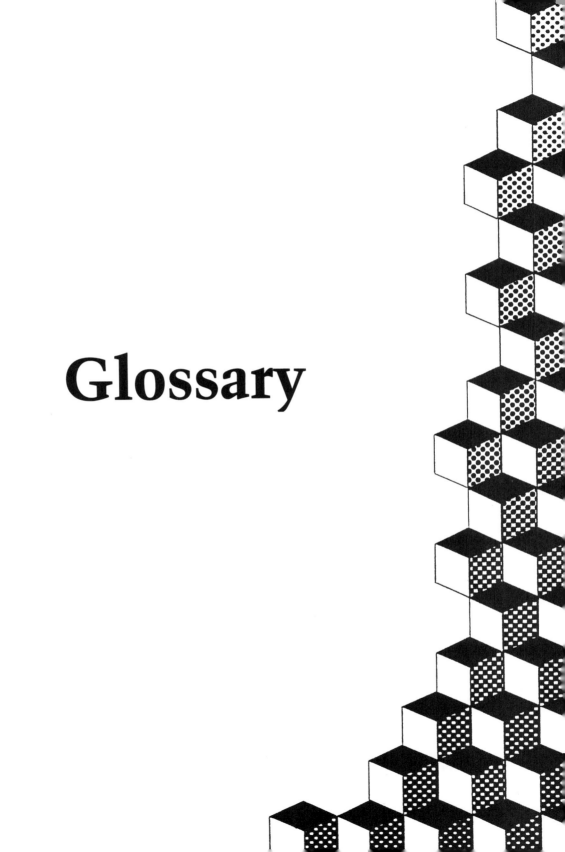

access time The amount of time it takes the computer to find and read data from a disk or from memory. The average access time for a hard disk is based on the time it takes the head to seek and find the specified track, the time for the head to lock onto it, and the time for the head to spin around until the desired sector is beneath the head.

active partition The partition on a hard disk that contains the boot and operating system. A single hard disk can be partitioned at the initial formatting of disk into several logical disks, such as drive C, drive D, and drive E. Only one partition, usually drive C, can contain the active partition.

adapter boards or cards The plug-in boards needed to drive monitors. Most monitor boards are monochrome graphic adapters (MGA), color graphic adapters (CGA), or enhanced graphic adapters (EGA). The EGA boards give a higher resolution than the CGA when used with a high resolution monitor. The video graphics adapters (VGA) can give an even higher resolution than the EGA.

algorithm A step-by-step procedure, scheme, formula, or method used to solve a problem or accomplish a task. It could also be a subroutine in a software program.

allocation unit When a disk is formatted, each track is divided into sectors. On the 360K and 720K, two sectors are considered to be an allocation unit. On the 1.2M and 1.44M, a single sector is considered an allocation unit. Some hard disks use four to 16 sectors as a single allocation unit. No two files or any part of two different files may be written in an allocation unit. The old term for allocation unit was *cluster*.

alphanumeric Data that has both numerals and letters.

analyst A person who determines the computer needs to accomplish a given task. The job of an analyst is similar to that of a consultant. Note that there are no standard qualifications requirements for either of these jobs; anyone can call themselves an analyst or a consultant. They should be experts in their field but might not be.

ANSI American National Standard Institute and a standard adopted by MS-DOS for cursor positioning. It is used in the ANSI.SYS file for Device drivers.

ASCII American Standard Code for Information Interchange. Binary numbers from 0 to 127 represent the upper- and lowercase alphabet letters, numbers 0–9, and several symbols found on a keyboard. A block of eight 0s and 1s are used to represent all of these characters. The first 32 characters, 0–31, are reserved for non-character functions of a keyboard, modem, printer, or other device. Number 32 (00100000) represents the blank space, which is a character. The numeral 1 is represented by the binary number for 49, which is 00110001. Text written in ASCII is displayed on the computer screen as standard text. Text written in other systems, such as WordStar, has several other characters added and is very difficult to read. Another 128 character representations have been added to the original 128 for graphics and programming purposes.

ASIC An acronym for Application Specific Integrated Circuit.

assembly language A low-level machine language made up of 0s and 1s.

asynchronous   A serial type of communication where one bit at a time is transmitted. The bits are usually sent in blocks of eight 0s and 1s.

AUTOEXEC.BAT   If present, this file is run automatically by DOS after it boots up. It is a file that you can configure to suit your own needs, to load and run certain programs, or to configure your system.

BASIC   Beginners All Purpose Symbolic Instruction Code, a high-level language that was once very popular. Many programs and games still use it. BASIC programs usually have a .BAS extension.

.BAK files   Any time that you edit or change a file in some of the word processors and other software programs, they will save the original file as a backup and append the extension .BAK to it.

batch   The batch command can be used to link commands and run them automatically. The batch commands can be made up easily by the user and all have the extension .BAT.

baud   A measurement of the speed or data transfer rate of a communications line between the computer and printer, modem, or another computer. Most present-day modems operate at 1200 baud, which is 1200 bits or about 120 characters per second.

benchmark   A standard-type program against which similar programs can be compared.

bidirectional   Both directions. Most printers print in both directions, thereby saving the time it takes to return to the other end of a line.

binary   Binary numbers are 0s and 1s.

BIOS   An acronym for Basic Input Output System. The BIOS is responsible for the control or transfer of data between and among the various peripherals.

bitmapped   The representation of a video image stored in the computer memory. Fonts for alphanumeric characters are usually stored as bit maps. When the letter A is typed, the computer goes to its library and pulls out a pre-formed A and sends it to the monitor. If a different size A (font) is needed, it will require another bit-map set. Graphic images can also be bit-mapped but consume an enormous amount of memory. Newer techniques allow different size and type of fonts to be scaled rather than bitmapped. *See* typeface.

bits   binary digits. A contraction of Binary and digITs.

boot or **bootstrap**   When a computer is turned on, all the memory and other internal operators must be set or configured. The IBM takes quite a while to boot up because it checks all the memory parity and most of the peripherals. A small amount of the program that does this is stored in ROM. Using this, the computer pulls itself up by its bootstraps. Sometimes you must do a *warm boot* (rebooting the computer by hitting Ctrl-Alt-Del) to get the computer out of an endless loop or some other hang- up.

buffer   A buffer is usually some discrete amount of memory used to hold data. A computer can send data thousands of times faster than a printer or modem can utilize it. In many cases, though, the computer can do nothing else until all of the data has been transferred. The data can be input to a buffer, which can then feed the data into the printer as needed. The computer is then freed to do other tasks.

bug and **debug** The early computers were made with high voltage vacuum tubes. It took rooms full of hot tubes to do the job that a credit card calculator can do today. One of the large systems went down one day. After several hours of troubleshooting, the technicians found a large bug that had crawled into the high voltage wiring. It had been electrocuted but subsequently had still managed to short out the whole system. Since that time, any type of trouble in a piece of software or hardware is called a *bug*. To debug a computer means to try to find all of the errors or defects.

bulletin boards Usually a computer with a hard disk that can be accessed with modem. Software and programs can be uploaded or left on the bulletin board by a caller, or a caller can scan the software that has been left there by others and download any that s/he likes. The BBs often have help and message services and are a great source of help for a beginner.

Burst Mode When the bus is taken over and a packet of data is sent as a single unit. During this time, the bus cannot be accessed by other requests until the burst operation is completed. This allows as much as 33M per second or more to be transmitted over the bus.

bus Wires or circuits that connect a number of devices together. It can also be a system. The IBM PC bus is the configuration of the circuits that connect the 62 pins of the 8 slots together on the motherboard. It has become the *de facto* standard for the clones and compatibles.

byte A byte is 8 bits, or a block of 8 0s and 1s. These 8 bits can be arranged in 256 different ways, figured out by taking the number of possibilities (2) to the power of the number of slots (8), which is 2 to the 8th power. Therefore, one byte can be made to represent any one of the 256 characters in the ASCII character set. It takes one byte to make a single character; because there are four characters in the word "byte," for example, it requires four bytes or 32 bits.

cache memory A buffer set up in high-speed memory to hold most frequently used data of a program being processed. Often a program will loop in and out of memory several times using the same data over and over. If it is stored in a fast on board cache, it can speed up the processing time considerably. The cache must be designed into the motherboard. *See* disk cache.

carriage width The width of a typewriter or printer. The two standard widths are 80 columns and 132 columns.

cell A place for a single unit of data in memory, or an address in a spreadsheet.

Centronics parallel port A system of 8-bit parallel transmission first used by the Centronics Company. It has become a standard and is the default method of printer output on the IBM.

character A letter, a number, or an 8-bit piece of data.

chip An integrated circuit, usually made from a silicon wafer. It can be microscopically etched and have thousands of transistors and semiconductors in a very small area. The 80286 CPU used in the AT has an internal main surface of about $1/2''$ and contains 120,000 transistors.

CISC An acronym for Complex Instruction Set Computing, which is the standard type of computer design as opposed to the RISC (Reduced Instruction Set Computers) ones used in larger systems. It might require as many as six steps for a CISC system to carry out a command, while the RISC system might need only two steps to perform a similar function.

clock The operations of a computer are based on very critical timing, so they use a crystal to control their internal clocks. The standard frequency for the PC and XT is 4.77 million cycles per second, or million Hertz. The turbo systems operate at 6 to 8MHz.

cluster Two or more sectors on a track of a disk. Each track of a floppy disk is divided into sectors. The term cluster is not used very often now, with its replacement term being *allocation unit*.

COM Usually refers to serial ports COM1 or COM2. These ports are used for serial printers, modems, a mouse or other pointing device, plotters, and other serial devices.

.COM A .COM or .EXE extension on the end of a filename indicates that it is a program that can run commands to execute programs.

COMMAND.COM An essential command that must be present in order to boot and start the computer.

COMDEX The nation's largest computer exposition and show, usually held once in the spring in Atlanta and in the fall in Las Vegas.

composite video A less expensive monitor that combines all the colors in a single input line.

console In the early days, a monitor and keyboard was usually set up at a desk-like console. This term has stuck, with a console nowadays referring to a computer. The command COPY CON allows you to use the keyboard as a typewriter. Type COPY CON PRN or COPY CON LPT1, and everything you type will be sent to the printer. At the end of your file or letter, type Ctrl-Z or F6 to stop sending.

consultant Someone who is supposed to be an expert who can advise and help you determine what your computer needs are. The consultant is similar to an analyst. No standard requirements or qualifications must be met to bear these titles, so anyone can call themselves an analyst or consultant.

conventional memory The first 640K of RAM memory, the memory that DOS handles. The PC actually has 1M of memory, but the 384K above the 640K is reserved for system use.

Co-Processor Usually an 8087, 80287, or 80387 that works in conjunction with the CPU and vastly speeds up some operations.

copy protection A system that prevents a diskette from being copied.

CPS When referring to a printer, the speed that it can print characters per second.

CPU Central Processing Unit such as the Intel 8088, 80286, 80386, or 80486.

CSMA/CD An acronym for Carrier Sense Multiple Access with Collision Detection. A network system that controls the transmissions from several nodes. It detects if two stations try to send at the same time and notifies the senders to try again at random times.

CRT A cathode ray tube, the large tube that is the screen of computer monitors and TVs.

current directory The directory in use at the time.

cursor The blinking spot on the screen that indicates where the next character will be input.

DATE command Date will be displayed anytime DATE is typed at the prompt sign.

Daisy Wheel A round printer or typewriter wheel with flexible fingers that have the alphabet and other formed characters.

database A collection of data, usually related in some way.

DES Data Encrytion Standard first developed by IBM. It can be used to encrypt data, making it almost impossible to decode it unless you have the code.

DIP An acronym for Dual In-line Package. It refers to the two rows of pins on the sides of most integrated circuit chips.

disk cache In processing some software, the program will request the same data from the disk over and over again. These requests can be quite time-consuming, depending on the access speed of the disk drive and the location of the data on the disk. In contrast, requested data cached in memory can be accessed almost immediately. *See* cache memory.

Disk Controller A plug-in board used to control the hard and/or floppy disk drives. All of the read and write signals go through the controller.

DMA Direct Memory Access. Some parts of the computer such as the disk drives can exchange data directly with the RAM without having to go through the CPU.

DPMI An acronym for DOS Protected Mode Interface. A proposed specification to govern the interaction of large applications with each other, DOS, and OS/2.

documentation Manuals, instructions, or specifications for a system, hardware, or software.

DOS Disk Operating System. Software that allows programs to interact and run on a computer.

dot matrix A type of printer that uses a matrix of thin wires or pins to make up the print head. Electronic solenoids pushed the pins out to form letters out of dots that were made when the pins pushed against the ribbon and paper. Older printers used seven pins, which gave rather poor quality print. Newer 24-pin heads can print in near letter quality (NLQ) type.

double density At one time, most diskettes were single-sided and had a capacity of 80 to 100K. Then the capacity was increased and technology was advanced so that the diskettes could be recorded on both sides with up to 200K per side double- sided, double density. Then quad density was soon introduced with 400K per side. Then, of course, the newer 1.6M high density diskettes came along.

All of these numbers are accurate before formatting. Most double-density disks are the common 360K formatted, with the quad ending up with 720K when formatted and the high density being 1.2M. The new 3.5" diskettes standard format will be 720K, while the high density 3.5" diskettes will hold 1.44M.

DRAM Dynamic Random Access Memory. This is the usual type of memory found in personal computers and is the least expensive of memory types.

DTP DeskTop Publishing, a rather loose term that can be applied to a small personal computer and a printer as well as to high powered sophisticated systems.

dumb terminal A terminal tied to a mainframe or one that does not have its own microprocessor.

duplex A characteristic of a communications channel that enables data to be transmitted in both directions. Full duplex allows the information to be transmitted in both directions simultaneously. In half duplex, it can be transmitted in both directions but not at the same time.

EATA An acronym for Enhanced AT Attachment, a standard proposed by the Common Access Method (CAM) committee. Their proposal would define a standard interface for connecting controllers to PCs. It would define a standard software protocol and hardware interface for disk controllers, SCSI host adapters, and for other intelligent chip embedded controllers.

echo A command that can cause information to be displayed on the screen from a .BAT or other file. Echo can be turned on or off.

EEPROM An Electrically Erasable Programmable Read-Only Memory chip.

EGA Enhanced Graphics Adapter. Used for high-resolution monitors.

E-mail Electronic mail, a system that allows messages to be sent through LANs (Local Area Networks) or by modem over telephone lines.

EMS Expanded Memory Specification, a specification for adding expanded memory put forth by Lotus, Intel, and Microsoft (LIM EMS).

EPROM An Erasable Programmable Read-Only Memory chip.

Ergonomics The study and science of how the human body can be most productive in working with machinery. This science includes the study of the effects of things like the type of monitor, the type of chair, lighting, and other environmental and physical factors.

errors DOS displays several error messages if it receives bad commands or problems of some sort exist.

ESDI(Enhanced System Device Interface) A hard disk interface that allows data to be transferred to and from the disk at a rate of 10 megabits per second. The older standard ST506 allowed only 5 megabits per second.

.EXE A file extension that indicates that the file is executable and can run and execute a program. Files with an .EXE extension are similar to .COM files.

expanded memory Memory that can be added to a PC, XT, or AT. It can only be accessed through special software.

expansion boards Boards that can be plugged into one of the 8 slots on the motherboard to add memory or other functions.

extended memory Memory that can be added to an 80286 or 80386 that will be addressable with the OS/2 operating system.

external commands DOS commands not loaded into memory when the computer is booted.

FAT An acronym for the File Application Table, which is a table on the disk that DOS uses to keep track of all of the parts of a file. A file could be placed in sector 3 of track 1, sectors 5 and 6 of track 10, and sector 4 of track 20. The File Application Table would keep track of where they are and will direct the read or record head to those areas.

FAX A shortened form of the word facsimile and X for transmission. A FAX machine scans an image or textual document and digitizes it in a graphical form. As it scans an image, a 0 or 1 is generated depending on the presence or absence of darkness or ink. The 0s and 1s are transmitted over the telephone line as voltages. *See* modem entry.

fonts The different types of print letters such as Gothic, Courier, Roman, Italic, and others. Each is a collection of unique characters and symbols. A typeface becomes a font when associated with a specific size.

format The process of preparing a disk so that it can be recorded. The format process lays down tracks and sectors so that data can be written anywhere on the disk and recovered easily.

fragmentation If a diskette has several records that have been changed several times, bits of files exist on several different tracks and sectors. This scattering of related data slows down writing and reading of the files because the head must move back and forth to the various tracks. If these files are copied to a newly formatted diskette, each file will be written to clean contiguous tracks, which will decrease the access time to the diskette or hard disk.

friction feed A printer that uses a roller or platen to pull through the paper.

game port An Input/Output (I/O) port for joysticks, trackballs, paddles, and other devices.

gigabyte One billion bytes. This will be a common size memory in a very short time. In virtual mode, the 80286 can address this much memory.

glitch An unexpected electrical spike or static disturbance that can cause loss of data.

global A character or something that appears throughout an entire document or program.

googol A very large figure—1 followed by 100 zeros.

GUI An acronym for Graphical User Interface. It usually makes use of a mouse, icons, and windows, such as those used by the Macintosh.

handshaking A protocol or routine between systems—usually the printer and the computer—to indicate readiness to communicate with each other.

hardware The physical parts that make up a computer system, such as disk drives, keyboards, monitors, etc.

hard disk A disk drive that can usually store a large amount of data. It has one or more magnetically coated platters that spin at 3600 RPM in a sealed casing.

Hayes-compatible Hayes was one of the first modem manufacturers. Like IBM, they created a set of standards that most others have adopted.

hexadecimal A system that uses the base 16. Our binary system is based on 2, while our decimal is based on 10. The hexadecimal goes from 00, 01, 02, 03, 04, 05, 06, 07, 08, 09, 0A, 0B, 0C, 0D, 0E, 0F. 10 would be 16 decimal, and the system starts over, so 20 would be 32 in decimal. Most of the memory locations are in hexadecimal notation.

hidden files Files that don't show up in a normal directory display, such as the DOS files necessary to boot a computer. They are hidden so that they will not be accidentally erased. Some programs such as XTree can be used to change the attributes so that the files can be erased or copied.

high level language A language such as BASIC, Pascal, or C. These program languages are fairly easy to read and understand.

ICs Integrated Circuits. The first integrated circuit was two transistors placed in a single can early in the 1960s. People found ways to put several semiconductors in a package, which was called SSI (Small Scale Integration). Then LSI (Large Scale Integration) and VLSI (Very Large Scale Integration) were developed. Today, we have VHSIC or Very High Scale Integrated Circuits (funny enough, we've almost run out of descriptive adjectives).

IDE An acronym for Integrated Disk Electronics. Western Digital and other companies are manufacturing hard drives with most of the controller circuitry on the disk assembly, but they still need an interface of some sort to connect to the computer. They are somewhat similar to SCSI and ESDI.

interface A piece of hardware or a set of rules that allows communications between two systems.

internal commands Those commands loaded into memory when DOS boots up.

interpreter A program that translates a high-level language into machine-readable code.

ISDN Integrated Services Network, a standard for telephone communications for transmission of voice, data, and images.

kilobyte 1000 bytes or 1K, or (more exactly) 1024 bytes. This number is 2 to the 10th power.

LAN An acronym for Local Area Network, where several computers might be tied together or to a central server.

laser printer A type of printer that uses the same type of "engine" used in copy machines. A laser beam electronically controlled sweeps across a drum, charging it with an image of the letters or graphics to be printed. The charged drum then picks up toner particles and deposits them on the page so that a whole page is printed at once.

LIM-EMS Acronym for the Lotus-Intel-Microsoft Expanded Memory Specification.

low-level format Most hard disks must have a preliminary low-level format performed on them before they can be formatted for DOS. Low-level formatting is also sometimes called *initializing*.

low-level language A machine-level language usually made of binary digits (which would be very difficult for the ordinary person to understand).

LQ Letter Quality, the type from a daisy wheel or formed-type printers.

macro A series of keystrokes that can be recorded, somewhat like a batch file, and then be typed back when one or more keys is pressed. For instance, I can type my entire return address with just two keystrokes.

mainframe A large computer that might serve several users.

megabyte 1,000,000 bytes, or 1M. More precisely, it's 2 to the 20th power (2^{20}), or 1,048,576 bytes. It takes a minimum of 20 data lines to address 1M, a minimum of 24 lines (2^{24}) to address 16M, and a minimum of 25 lines (2^{25}) to address 32M.

menu A list of choices or options. A menu-driven system makes it very easy for beginners to choose what they want to run or do.

MFM Modified Frequency Modulation, the scheme for the standard method of recording on hard disks. *See* RLL.

MHz Megahertz, a million cycles per second. Older technicians still call it CPS. A few years ago, a committee decided to honor Heinrich Rudolf Hertz, (1857–1894), for his early work in electromagnetism, so they changed the cycles per second (CPS) to Hertz or Hz.

mode A DOS command that must be invoked to direct the computer output to a serial printer.

modem An acronym for MOdulator-DEModulator. A device that allows data to be sent over telephone lines. A modem creates digital voltages that are changed (modulated) to analog voltage while it is being transmitted over the telephone lines and then demodulated by the receiving modem.

modes The 80286 and 80386 will operate in three different modes—the real, the protected and the virtual. For more details, see Chapter 9.

mouse A small pointing device that can control the cursor and move it anywhere on the screen. It usually has two or three buttons that can be assigned various functions.

MTBF An acronym for Mean Time Before Failure, which is an average of the time between failures, usually used in describing a hard disk or other component.

multitasking The ability of the computer to perform more than one task at a time. Many of the newer computers have this capability when used with the proper software.

multiuser A computer capable of providing service to more than one user, such as a server for a local area network (LAN).

NEAT Chipset An acronym describing the New Enhanced AT chipset from Chips and Technology. Chips and Technology combined the functions of several chips found on the original IBM motherboard into just a few very large scale integrated circuits (VLSI). These chips are used on the vast majority of clone boards.

NLQ An Acronym for Near Letter Quality, the better-formed characters from a dot matrix printer.

null modem cable A cable with certain pairs of wires crossed over. If the computer sends data from pin 2, the modem might receive it on pin 3. The modem would send data back to the computer from its pin 2 and be received by the computer on pin 3. Several other wires would also be crossed.

OOP Object-Oriented Programs, a type of programming that utilizes parts of existing programs to provide new applications.

OS/2 An operating system allowing the 80286 and 80386 machines to directly address huge amounts of memory. It removes many of the limitations that DOS imposes. OS/2 will not benefit the PCs or XTs to any great degree.

parallel A system that uses 8 lines to send 8 bits (one whole byte) at a time.

parity checking In the computer memory system, an error detection technique that verifies the integrity of the RAM memory contents. This is the function of the ninth chip in a memory bank. Parity checking systems are also used in other areas such as verifying the integrity of data transmitted by modem.

plotter An X-Y writing device that be used for charts, graphics, and many other functions that most printers can't handle.

prompt The sign showing that DOS is waiting for an entry.

protocol The rules and methods by which computers and modems can communicate with each other.

QIC Quarter Inch Cartridge tape, a width of tape used in tape backup systems. Some standards using this size tape have been developed, but several non-standard systems are still in use.

RAM An acronym for Random Access Memory, which is computer memory used to temporarily hold files and data as they are being worked on, changed, or altered. It can be written to and read from. It is a *volatile* memory, meaning that any data stored in it is lost when the power is turned off.

RGB For Red, Green, and Blue, the three primary colors used in color monitors and TVs. Each color has its own electron gun that shoots streams of electrons to the back of the monitor display and causes it to light up in the various colors.

RISC An acronym for Reduced Instruction Set Computing. A design that allows a computer to operate with fewer instructions, thus letting it run much faster.

RLL Run Length Limited. A scheme of hard disk recording that allows 50% more data to be recorded on a hard disk than the standard MFM scheme. ARLL or ERLL (for Advanced and Enhanced RLL) will allow twice as much data to be recorded on a hard disk. The older MFM system divided each track into 17 sectors of 512 bytes each, the RLL format divides the tracks into 26 sectors with 512 bytes each, and the ARLL and ERLL divides them into 34 sectors per track.

ROM Read-Only Memory. It does not change when the power is turned off. The primary use of ROM is in the system BIOS and on some plug-in boards.

scalable typeface Unlike bitmapped systems, where each font has one size and characteristic, scalable systems allow typeface to be shrunk or enlarged to different sizes to meet specific needs. Once a font has been scaled, it is then stored as bitmapped, allowing much more flexibility and using less memory. There are also scalable graphic systems.

SCSI An acronym for Small Computer System Interface, pronounced "scuzzy," which is a fast parallel hard disk interface system developed by Shugart Associates and adopted by the American National Standards Institute (ANSI). The SCSI system allows multiple drives to be connected, supporting a transfer rate of 1.2M per second. Because a byte is 8 bits, this is about the same as the ESDI 10 megabit per second rate.

sector A section of a track on a disk or diskette. A sector ordinarily holds 512 bytes. A 360K disk has 40 tracks per side, and each track is divided into 9 sectors. Each track of the 1.2M disk is divided into 15 sectors (18 sectors on the 1.44M). The tracks on a hard disk might be divided into as many as 56 sectors.

serial The transmission of one bit at a time over a single line.

SIMM An acronym for single inline memory module for DRAMs.

SIP An acronym for single inline package. It can be a form of DRAM memory, small resistor packs, or integrated circuits. They have a single line of pins.

source The origin or diskette to be copied from.

SPARC An acronym for Scalable Processor Architecture, a RISC system developed by Sun Microsystems for workstations.

spool An acronym for Simultaneous Peripheral Operations On Line. A spooler acts as a storage buffer for data, which is then fed out to a printer or other device. In the meantime, the computer can be used for other tasks.

SRAM Static RAM, a type of RAM that can be much faster than DRAM. SRAM is made up of actual transistors turned on or off and maintaining their state without constant refreshing such as needed in DRAM. SRAM is considerably more expensive and requires more space than DRAM.

target The diskette to be copied to.

time stamp The record of the time and date recorded in the directory when a file is created or changed.

tractor A printer device with sprockets or spikes that pulls the computer paper with the holes in the margins through the printer at a very precise feed rate. A friction-feed platen might allow the paper to slip, move to one side or the other, or not be precise in the spacing between the lines.

Trojan horse A harmful piece of code or software, usually hidden in a software package, that will later cause destruction. It differs from a virus in that it does not grow and spread.

TSR A acronym for Terminate and Stay Resident. When a program such as Sidekick is loaded in memory, it will normally stay there until the computer is booted up again. If several TSR programs are loaded in memory, not enough memory might be left to run some programs.

turbo Usually means a computer with a faster-than-normal speed.

user friendly Usually means bigger and more expensive. It should make using the computer easier. Memory is now less expensive, so large programs are being developed to use more memory than ever before.

user groups Usually a club or a group of people who use computers. Often the club will be devoted to users of a certain type of computer, but most clubs welcome anyone to join.

vaporware Products that are announced (usually with great fanfare) but aren't yet ready for market.

VCPI An acronym for Virtual Control Program Interface, a plug-in board that allows a computer to address memory above 640K.

virtual Something that might be essentially present but not in actual fact. If you have a single disk drive, it will be drive A but you also have a virtual drive B.

virus Destructive code placed or embedded in a computer program, usually created by a sick person who just wants to hurt others. The virus is usually self-replicating and will often copy itself onto other programs. It might remain dormant for some time and then completely erase a person's hard disk.

volatile Refers to memory units that lose stored information when power is lost. Non-volatile memory would be that of a hard disk or tape.

VRAM Video RAM, a type of special RAM used on video or monitor adapters. The better adapters have more memory so that they can retain full-screen high-resolution images.

windows Many new software packages are now loaded into memory, stay in the background until called, and then pop up on the screen in a window. The Microsoft company has a software package called Windows that provides an operating environment for many DOS programs.

Index

A
accelerator boards, 54-55
address labels, laser printer, 145-146

B
backup, 87-97
   advantages, 92-93
   BACKUP.COM, 93-94
   Digital Audio Tape (DAT), 95
   excuses, 90-91
   external plug-in hard disk drives, 97
   file-oriented, 93
   hard disk cards, 97
   image, 93
   methods, 93-97
   second hard disk drive, 96-97
   software, 93-94
   tape systems, 94-95
   unerase software, 88-89
   very high density disk drives, 96
   videotape, 96
battery cable, installing, 23-24
baud rate, 153
Baudot, Emile, 153
Bernoulli, Jakob, 69
Bernoulli disk drives, 69
boards
   accelerator, 54-55
   CPU daughter, 55
   multifunction, 18

bulletin boards, 156-157
   cost, 157-158
   illegal activities, 157
   locating, 158
   software viruses/Trojan horses, 157

C
cables, monitor, 108
cache system, 10
case, 15, 44-45
   cost, 45
Cathode Ray Tube (CRT), 102
CD-ROM, 83-85
   High Sierra standard, 84-85
clock circuits, 75
Color Graphics Adapter (CGA), 104, 108
components
   connected, 33, 36
   cost, 8-9
   installing into case, 36-40
   necessary, 7-20
   troubleshooting, 198
computer (*see also* 80386 CPU)
   accessories (*see* specific types; i.e., disk drives, keyboards, monitors, etc.)
   assembly instructions, 23-26
   assembly instructions for IBM PC-XT upgrade, 46-54
   building vs. buying, 2-3
   building vs. upgrading, 3

   components, 7-20
   cost, 3
   cost to build, xv
   IBM compatibility, xvii
   necessity, 1-6
   system cost, 15
   testing completed system, 40-41
   upgrading, 43-55
   upgrading, cost, 44-45
   upgrading, IBM PC-XT, 43-55
   upgrading, IBM PC/XT/AT, xviii
Computer Aided Design (CAD), 181
Continuous Edge Graphics (CEG), 105
controller, floppy disk, 68
controller cables
   connecting to boards, 27-28
   connecting to floppy disk, 27
   connecting to floppy disk drive A, 33
   connecting to hard disk drive, 27, 29, 52
controllers, hard disk, 78
coprocessors, 10
   cost, 12

D
daisy-wheel printers, 140
data cable, connecting to hard disk drive, 30, 52

223

daughter boards, 55
Digital Audio Tape (DAT), 95
digitizers, 134-135
disk controllers, 19
disk drives, 19 (*see also* floppy disk drives; hard disk drives)
   attaching plastic side rails, 27-29
dot-matrix printers, 138-140
   advantages, 139
   color, 139
   cost, 140
   noise reduction, 139-140
   speed, 139
DRAM, 114-117
   cache memory, 116
   interleaved memory, 116
   refreshment/wait states, 115-116
drivers, 105-108, 110-111
Dual In-line Package (DIP), 117
Dynamic Data Exchange (DDE), 166
Dynamic RAM (*see* DRAM)

E

E-mail, 158-159
80386 CPU
   applications, xv-xvii
   assembly instructions, 23-36
   benchtop assembly, 22-36
   cache system, 10
   coprocessors, 10-12
   cost, 22
   EISA, 12-14
   enhanced mode, 122
   features, 5
   ISA, 12-13
   microprocessor chip, 4-5
   operating frequencies, 10
   system cost, 15
   testing completed system, 40-41
   vs. other CPUs, 4-5
80386DX CPU
   assembly, 21-42
   vs. 80386SX CPU, xv
80386DX motherboard, 46-47
80386SX CPU
   assembly, 21-42
   vs. 80386DX CPU, xv
80386SX motherboard, 9-19
EISA, 12-14
   connector, 13-14
   gang of nine, 13
   necessity, 14
Electrostatic Discharge (ESD), 202
Enhanced Graphics Adapter (EGA), 104, 108
Enhanced Small Device Interface (ESDI), 78
Extended Industry Standard Architecture (*see* EISA)
external hard disk drives, 81
eXtra Graphics Array (XGA), 104

F

facsimile boards, 159-162
   installing, 162
facsimile machines, 159-162
   mail order FaxBack, 191-192
File Allocation Table (FAT), 59-60
   jumbled, 89
file-oriented backup, 93
fixed disk drives (*see* hard disk drives)
floppy disk drives, 57-70
   Bernoulli, 69
   connecting 3.5-inch drive to board, 49
   connecting controller cable, 27
   connecting controller cable to drive A, 33
   connecting middle connector to 3.5 inch, 30, 32
   connecting power cable, 50-51
   connecting power cable adapter to 3.5 inch, 30-31
   connecting power cable to 3.5 inch, 30, 32
   connecting power cable to drive A, 33-34
   extended density, 68
   high density systems, 68-69
   mounting 3.5 inch, 69
   very high density, 68-69
   what to buy, 69-70
floppy disks
   allocation units, 59
   basics, 59-62
   controllers, 68
   cost, 67-68
   cylinders, 60-61
   directory limitations, 60
   discount sources, 68
   FAT, 59-60
   5.25-inch, 62-64
   formatting, 59, 65-67
   heads, 61
   read accuracy, 61
   rotation speed, 62
   sectors, 59
   TPI, 61
   3.5-inch, 62-65
      tracks, 59
   write protection, 88
formatting
   .BAT files, 66
   converting 720K to 1.44M disk, 66-67
   floppy disks, 59, 65-67

G

glossary, 209-221
Graphical User Interface (GUI), 164
graphics tablets, 134-135

H

hard disk cards, 80-81, 97
hard disk drives, 19, 71-85, 97
   adding space, 81-82
   aligning one or more, 80
   capacity, 73
   choosing, 72-73
   CMOS ROM setup, 80
   compression, 81-82
   connecting controller cable, 27, 29, 52
   connecting data cable, 30, 52
   connecting data cable to controller card, 24, 26
   connecting power cable, 30-31, 53
   controllers, 78-79
   determining size needed, 74
   ESDI system, 78
   external, 81, 97
   formatting, 79-80
   head crash, 89-90
   head crash recovery, 90
   head positioners, 76
   history, 73-74
   IDE system, 77
   installing controller card, 25
   leaving on, 74
   logical drives, 90
   MFM system, 76-77
   MTBF, 75
   operation, 74-76
   platters, 75-76
   reading/writing, 75

RLL system, 77
SCSI system, 78
second for backup, 96-97
speed/access time, 73
stepper motor system, 73
systems, 76-79
types, 73
voice coil system, 73
hardware, problems, 207
Hercules Monographic Adapter (HMGA), 108
high resolution programs, 105-108

I

image backup, 93
Industry Standard Architecture (ISA), 12-13
ink jet printers, 140-141
color, 141
input devices, 123-136
Integrated Drive Electronics (IDE) system, 77
Integrated Services Digital Network (ISDN), 159

K

keyboards, 19, 45, 124-131
connecting cable, 33, 35
cost, 19
covers, 124
keys, 126-130
model switch, 126
operation, 126-130
sources, 130-131
specialized, 131
standards, 124-126
trackball combination, 134
keys, 126-130
*, 128
Alt, 130
arrow, 127
backspace, 129
break, 129
CapsLock, 129
Ctrl, 129
Del, 128
DOS, 127
End, 128
Enter, 129
Esc, 128
function, 126-127
function, reprogramming, 130
Home, 128
Ins, 128
number, 127
PgDn, 128
PgUp, 128
PrtScr, 128
Return, 129
Scroll Lock, 128
Shift, 129
special, 126
special key functions, 130
Tab, 129

L

laser printers, 141-146
address labels, 145-146
cost, 142-143
engine, 141-142
extras, 143
fonts/scalable fonts, 143
maintenance, 145
memory, 143
paper, 145
paper size, 145
PDL, 143
resolution, 144-145
speed, 144

M

magazines, 189-194
computer, 189-190
FaxBack, 191-192
free to qualified subscribers, 190
mail order, 185-194
books, 194
Federal Trade Commission rules, 188-189
magazines, 189-194
rules, 187-188
make directory (MD), 60
mass storage, 71-85
CD-ROM, 83-85
compression, 81-82
High Sierra standard, 84-85
Mean-Time Before Failure (MTBF), 75, 90
memory, 113-122
cache, 116
conventional, 119
DRAM, 114-117
expanded, 120
explanation, 115
extended, 119-120
flash, 117
interleaved, 116
modes, 5-6, 120-122
motherboard, 117-118
RAM, 114
requirements, 119
ROM, 114
SRAM, 116-117
types, 119-120
Micro Channel Architecture (MCA), 12
modems
banking, 159
cables, 156
communications software, 152-153
external, 152
installing, 155-156
internal, 152
plug-in telephone line, 155-156
setting configuration, 155
testing, 156
types, 152-156
what to buy, 154-155
Modified Frequency Modulation (MFM), 76-77
monitors, 18, 99-112
adaptors, 18, 108-111
analog vs. digital, 110
bandwidth, 106
basic information, 101-105
cables, 108
CEG, 105
CGA, 104
cleaning screen, 107
color, 103
connecting cable, 33, 35
controlling the beam, 103
controls, 107
cost, 100, 107
dot pitch, 106
drivers, 105-108, 110-111
EGA, 104
glare, 107
glare shields, 107
high resolution programs, 105-108
interlaced vs. noninterlaced, 105-106
landscape vs. portrait, 106
monochrome, 103
pixels, 104
resolution, 104
scan rates, 102-103
screen size, 106-107
SEGA, 104
shopping checklist, 101
SVGA, 104

monitors (cont.)
  tilt-and-swivel base, 107
  VGA, 104
  VRAM, 110
  what to buy, 100-101, 111-112
  XGA, 104
Monochrome Display Adaptor (MDA), 108
motherboards, 9-19
  baby 80386DX, 46-47
  baby-size, 44
  cost, xiv, 10
  EISA, 12-14
  installing, 36, 38-39
  installing connectors for power 24-25
  ISA, 12-13
  sliding into case, 38-39
motherboards, memory, 117-118
  DIP, 117
  SIMM, 117
  SIP, 117-118
mouse
  cost, 134
  interfaces, 133
  systems, 132
  types, 132
multifunction boards, 18

O

on-line services, 158
Optical Character Readers (OCRs), 135-136

P

Page Description Languages (PDL), 143
parts, locating, 19-20
Perstor hard disk controller card, installing, 25
pixels, 104
platters, 75-76
plotters, 146-147
  installing, 147-148
  supplies, 147
port cables, installing, 23
PostScript printers, 144
power cable
  connecting to 3.5-inch floppy disk drive, 30, 32
  connecting to floppy disk drive, 50-51
  connecting to floppy disk drive A, 33-34

connecting to hard disk drive, 30-31, 53
showing correct installation, 48-49
power cable adapter, connecting to 3.5-inch floppy disk drive, 30-31
Power On Self Test (POST), 199-200
power supply, 16-18, 45
  connecting to motherboard, 48
  installing in case, 36-37
  troubleshooting, 200
printers, 137-149
  buffer, 140
  choosing, 138-146
  color, 146
  daisy-wheel, 140
  dot-matrix, 138-140
  ink jet, 140-141
  installing, 147-148
  laser, 141-146
  plotters, 146-147
  PostScript, 144
  sharing, 148
  sources, 149
protected mode, 6, 121-122
protocols, 153

R

RAM memory sockets, 18
Random Access Memory (RAM), 114
Read-Only Memory (ROM), 114
real mode, 5, 120-122
resolution, 104
ribbon cable, connecting to controller card, 24-26
Run Length Limited (RLL) system, 77

S

saving (see backup)
scanners, 135-136
  hand-held, 135-136
  OCR, 135
shareware, 159
  sources, 193
Single In-line Memory Module (SIMM), 117
Single In-line Package (SIP), 117
slot covers, 41-42
Small Computer Systems Interface (SCSI), 78

software
  ACT!, 184
  AMI, 175
  Andrew Tobias TaxCut, 182
  askSam, 176
  AutoCAd, 181
  backup, 93-94
  CheckIt, 179
  communications, 152-153
  computer-aided design (CAD) programs, 181
  Concurrent DOS 386, 174
  Crosstalk for Windows, 169
  database programs, 176-177
  dBASE IV, 176
  DesignCAD 2D, 181
  DesignCAD 3D, 181
  DESQview, 174
  diagnostic, 204-205
  directory programs, 180
  disk management programs, 180
  disk operating systems (DOS), 172-174
  Disk Technician, 179
  DOS help programs, 174
  DR DOS 5.0, 173
  Dragnet and Prompt, 169
  essential, 171-184
  Form Express, 184
  FoxPro, 177
  GeoWorks, 170
  grammar checkers, 175-176
  Grammartik, 176
  It's Legal, 183
  J.K. Lasser's Your Income Tax, 182
  Mace Utilities, 179
  Magellan 2.0, 181
  Micrografx, 169
  Microsoft Excel, 178
  Microsoft Word for Windows, 175
  miscellaneous, 183-184
  Money Counts, 183
  MS-DOS 5.0, 172-173
  Norton Desktop for Windows, 169
  Norton Utilities, 179
  OPTune, 179
  OS/2, 173-174
  packages used with Windows 3.0, 168-170
  PageMaker, 169

Paradox, 177
PC Tools, 179
PC-Write, 175
problems, 206
public domain, 159, 193
QDOS 3.0, 180
Quattro, 178
R:BASE 3.1, 176-177
Random House Encyclopedia, 183
Right Writer, 176
search utilities, 180-181
shareware, 159, 193
SideKick Plus, 180
SpinRite II, 179
spreadsheet programs, 177-178
SuperCalc5, 178
SwifTax, 182
tax programs, 181-183
TaxView, 182-183
Tree86 3.0, 180
Trojan horses, 157
TurboTax, 183
unerase, 88-89
utility programs, 178-180, 204-205
Ventura Publisher, 169
viruses, 157
WillMaker 4.0, 183
Windows 3.0, 163-170
Windows Express, 169
Wonder Plus 3.08, 180
word processing programs, 174-175
word processing with Windows, 170
WordPerfect, 175
WordStar, 174-175
XTreePro Gold, 180
YourWay, 168

sources, 6
standard mode, 121-122
Static Ram (SRAM), 116-117
Super Enhanced Graphics Adapter (SEGA), 104
Super Video Graphics Array (SVGA), 104
switch panel wires, 15-16, 40

T

telecommunications, 151-162
  banking, 159
  baud rate, 153-154
  bulletin boards, 156-157
  E-mail, 158-159
  estimating connect time, 154
  facsimile boards, 159-162
  facsimile machines, 159-162
  ISDN, 159
  modems, 152-156
  on-line services, 158
  protocols, 153
  public domain software, 159
  shareware, 159
  software, 152-153
  software viruses/Trojan horses, 157
  sources, 162
telecommuting, 162
television (*see* monitors)
TEST.BAT, 40-41
386 (*see* 80386 CPU)
tools, 201-203
trackballs, 134
  keyboard combination, 134
tracks per inch (TPI), 61
troubleshooting, 195-208
  common problems, 202
  components, 198
  dead computer, 205-206

diagnostic/utility software, 204-205
documenting problems, 196-197
electricity/electronics, 197
Electrostatic Discharge (ESD), 202
hardware problems, 207
instruments/tools, 201-202
levels, 197
locating problem, 203-204
Power On Self Test (POST), 199-200
power supply, 200-201
software problems, 206

U

uninterruptible power supply, 16-18

V

very large scale integration (VLSI), 44
Video Electronics Standards Association (VESA), 104
Video Graphics Array (VGA), 104, 109
Video RAM (VRAM), 110

W

Windows 3.0, 163-170
  applications, 168-170
  automatic setup, 165
  breaking the 640K barrier, 166
  features, 165-168
  installation, 165
  on-line help, 165-166
  operational modes, 166
  requirements, 165
  vs. OS/2, 168